AQUATIC FLORA

of

Connecticut & Rhode Island

Field Guide to Floating, Submersed & Floating-leaved Plants

DONALD J. PADGETT

AQUATIC FLORA OF CONNECTICUT & RHODE ISLAND
Field Guide to Floating, Submersed & Floating-leaved Plants

Donald J. Padgett

Spatterdock Press
Middleboro, Massachusetts

Printed by KDP, An Amazon.com Company

ISBN-13: 978-0-578-32522-4

Cover photo: Washington Co., RI © Michael Graziano

"So lovely was the loneliness of a wild lake"

Edgar Allan Poe, 1827

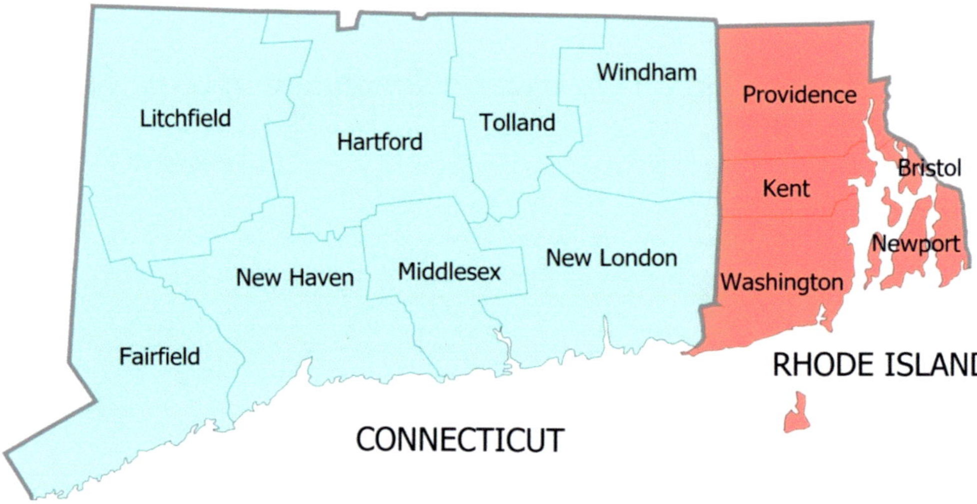
Litchfield
Hartford
Tolland
Windham
Providence
Bristol
Kent
Newport
New London
Middlesex
New Haven
Washington
Fairfield
RHODE ISLAND
CONNECTICUT

CONTENTS

PREFACE

The impetus for this book came from teaching courses on aquatic plant biology with the frustration of finding an appropriate field manual for students. While there are excellent field guides on aquatic plant identification for this region, most are either too advanced or far too simple, and always incorporate too broad a geographic coverage which unnecessarily adds many species to the mix for consideration. This book was constructed as a quasi-technical—yet comprehensive—identification resource devoted to the true aquatic macrophytes of Connecticut & Rhode Island. It provides taxonomic keys, illustrations, color photographs, basic distribution maps, and general accounts of each species. I hope that it will be of use to all those with a professional or amateur interest in the aquatic plants of Connecticut or Rhode Island and surrounding areas.

DJP

ACKNOWLEDGEMENTS

I am always grateful to those who helped advance my knowledge of aquatic plants, particularly Drs. Garrett Crow, C. Barre Hellquist, Don Les, and Tom Philbrick. I extend gratitude to those who granted me use of their photographs or figures, especially Katy Chayka and Peter Dziuk of *Minnesota Wildflowers* (www.minnesotawildflowers.info), Barre Hellquist, Mike Graziano, and Hyun Jung Cho of *Mississippi Aquatic Plants Website* (www.jcho.masgc.org). As always, I am thankful to Bridgewater State University for continued support. Lastly, a note of love to Shannon, Morgan, Sullivan, and my canine companion Ali.

INTRODUCTION

Aquatic Flora of Connecticut & Rhode Island summarizes the distribution, habitat, and biology of plants that grow in fresh, brackish, or saline waters of the Constitution and Ocean States. It is intended to foster an appreciation of this fascinating group of plants and assist users in identifying them to species. This guide aims to update and bridge the gap between the more technical floras and the overly simple field guides of our region, with a focus on the floating, submersed, and floating-leaved flora. Consequently, advanced botanical terminology is used at a minimum, with simpler descriptive words or phrases used as a substitute.

I included all truly aquatic plants considered native or introduced in Connecticut and Rhode Island that one may encounter in ponds, lakes, pools, rivers, streams, and marshes. Truly aquatic species are defined here as those with most of their cycle spent floating on, or submerged in, water. For the sake of brevity, I included neither the numerous "emergent" wetland species—unless such species can also be found as a distinctive submersed growth form (e.g., *Gratiola aurea*)—nor plants that are interspecific hybrids (e.g., *Potamogeton* hybrids). Native or indigenous plants are those which occur naturally in Connecticut or Rhode Island. Introduced species entered the flora from other geographic regions either by intentional or accidental means. To make the book as comprehensive as possible, both common and rare species of angiosperms, ferns, and fern-allies are included. While less likely to be encountered in the field, the rare and endangered species were combined with the more common ones so that their place in the region's rich aquatic diversity may be fully appreciated, and with the hope that new occurrences may be discovered and protected.

While 139 species are described in detail, 20 additional species are discussed for comparison purposes or represented in various ways.

HOW TO USE THIS BOOK

Format. The book is arranged into an easy-to-use format. If the common name (e.g., Eurasian water-milfoil) of a plant is known, a quick check in the index can direct you to the page number for the species account. If the scientific generic (genus) name is known (e.g., *Myriophyllum*), flip through the descriptive text pages to locate the page for that plant(s). The descriptive pages for plants are arranged alphabetically by genus name, and then by species within the genus. Another method would be to thumb through the book and look for an illustration or photo of a plant like the one you are trying to identify. If the plant in question is completely unknown to you, or if you are not completely certain of its identity using the preceding methods, the plant can be identified using the taxonomic keys in the book.

Use of keys. Identification keys were constructed to facilitate the identification of species. They consist of a series of numbered couplets, with a choice of contrasting statements. After a choice is made, a number at the end of the statement directs you to either to the next couplet or a plant name. Two sets of keys are provided to help sort out this large group of plants: 1) keys to genera, and 2) keys to species within a genus. Once a genus is identified in the key, locate the genus descriptive page. If there are more than one species in the genus, each descriptive entry begins with an overview of the group and ends with a key to the various species.

Plant names. The scientific plant names in this book agree with *Flora Novae Angliae* (Haines 2011). To make the book useful to a wide audience, the plants are referred to by common names in addition to their scientific names (except in keys). If a species has more than one popular common name, alternate common names are provided within the text.

Species descriptions. At the top of each species descriptive page the accepted **scientific name** and author (who named the species) and **common name** is given. The family name in scientific form is included if the species is the only one for the genus and there is no genus descriptive page.

A description of the species is then provided with features useful for its identification emphasized. Descriptions are organized with a sequence of plant type, stems, leaves, flower clusters, flowers, and then fruits, and are based on the feature's appearance during the active growing season. If a feature is not especially useful for identifying a plant (e.g., anther number), it is not mentioned. **Habitat & Distribution** indicates native status, provides the common habitat types the species occupies, and its general range with an attempt to characterize its abundance in CT and RI. If the species is listed as rare or prohibited in a state, this information is noted here. **Notes** provides information regarding any specific environmental conditions the plant requires or offers, its role in a community, and similarity to other plant species.

A **map** shows the distribution of each species as documented within the counties of Connecticut and Rhode Island and includes known introductions or invasions. A county is shaded in which the species is known to occur. The maps differentiate by color between native (blue) and introduced (red) occurrences.

Illustrations and photographs. Each species account is supplemented with photographs and or illustrations. Attributable credits of each illustration or photo are found on Page 200.

AQUATIC PLANTS OF CT & RI

Composition

Connecticut and Rhode Island are each water-rich states and collectively contain more than 2600 lakes, 180 rivers, and estuarine waters which host a diverse assemblage of aquatic plants. The 139 hydrophytes considered in this book include three of the four plant life forms associated with aquatic habitats comprising 132 flowering plants (72 monocotyledons, 60 dicotyledons), 5 quillworts, and 2 ferns. These "true", or strictly aquatic, hydrophytes range from plants that grow completely submerged to floating-leaved and free-floating species. Most species (87%) are native, but 18 are considered non-native. Plants that may grow with their roots and lower stems under water but hold a more substantial amount of their body in the air—so called *emergent* species—are important components of a state's flora but omitted from this guide for brevity, as with lower plants, like aquatic liverworts, mosses, and Charophyte algae. Those few emergent species (i.e., *Gratiola aurea, Hypericum boreale, Pontederia cordata*) that also possess a distinctive submersed growth form, however, are included.

While most of these aquatic plants may be commonly encountered, 42 species are classified as endangered, threatened, or of concern by either Connecticut's Department of Energy and Environmental Protection or Rhode Island's Department of Environmental Management. Individuals of listed species are protected under provisions of the Connecticut Endangered Species Act (C.G.S.A. § 26-303 - 316) or Rhode Island State Endangered Species Act (R.I. Gen. Laws, 1956, § 20-37-1 to 5) from take (i.e., collect, destroy, harm) including buying, selling, and or transport. An additional three species are believed to be extirpated from Connecticut and absent in Rhode Island. These "historic" species (*Heteranthera reniformis, Nuphar advena,* and *Nuphar microphylla*) are still included in this guide since extant populations are known in other nearby states.

Most non-native aquatic plants introduced into Connecticut and Rhode Island are deemed invasive and banned (in CT) from sale or distribution. Invasive aquatic plants have impacted many lakes, ponds, reservoirs (>50% of surveyed lakes in CT; >100 lakes in RI) and rivers and streams (27 rivers surveyed in RI have invasives). Exotic species like *Hydrilla verticillata, Myriophyllum aquaticum, M. spicatum, M. heterophyllum, Najas minor, Potamogeton crispus,* and *Trapa natans* are infamous weeds that threaten native species and ecosystems.

Geographic Distribution

The natural occurrence of certain hydrophytes is largely regulated by chemical properties of their watery environment, as well as water depths and substrate preferences. Like all aquatic organisms, the optimal growth and survival of hydrophytes is influenced by levels of pH, alkalinity, and interactions of elements in the water. As for southern New England, the chemical composition of surface waters shows strong regional patterns related to the underlying geology (Canavan and Siver 1994; Mattson et al. 1992). Like what is observed in Massachusetts to the north, there is a general decline in pH and alkalinity in Connecticut waters from west to east as bedrock shifts from a prevalence of limestone and marble to granite and sand (Bell 1985; Figure 1). This pattern is reflected, to a degree, in the geographic distribution of many aquatic plants.

The high alkalinity, pH, and calcium levels characteristic of western Connecticut (particularly the Marble and Central Valleys) support species otherwise rarely found more eastwardly and absent from Rhode Island, like *Heteranthera dubia, Myriophyllum sibiricum, Sagittaria cuneata,* and select species of *Potamogeton* (*P. friesii, P. hillii, P. richardsonii, P. strictifolius*). The hard waters cause higher available nutrients, so these waterbodies tend to have lower water quality and clarity. Conversely, more highly acidic waters to the east and near the coast support species like *Myriophyllum pinnatum, Potamogeton confervoides, Proserpinaca pectinata, Sagittaria teres,* and *Utricularia subulata.* These soft waters are nutrient-poor and have high clarity.

Watercourses influenced by ocean tides, dominated by Na and Cl, support *Crassula aquatica, Eriocaulon parkeri, Isoetes riparia, Lilaeopsis chinensis, Limosella australis, Sagittaria montevidensis,* and *Zannichellia palustris* in brackish conditions, and *Ruppia maritima* and *Zostera marina* in more saline conditions. Of course, many hydrophytes are more broadly tolerant of water chemistry. This is reflected in their more widespread distributions (e.g., *Elodea nuttallii, Nuphar variegata, Nymphaea odorata, Potamogeton berchtoldii, P. epihydrus, P. spirillus, P. natans, P. amplifolius, Utricularia vulgaris,* and *U. purpurea*).

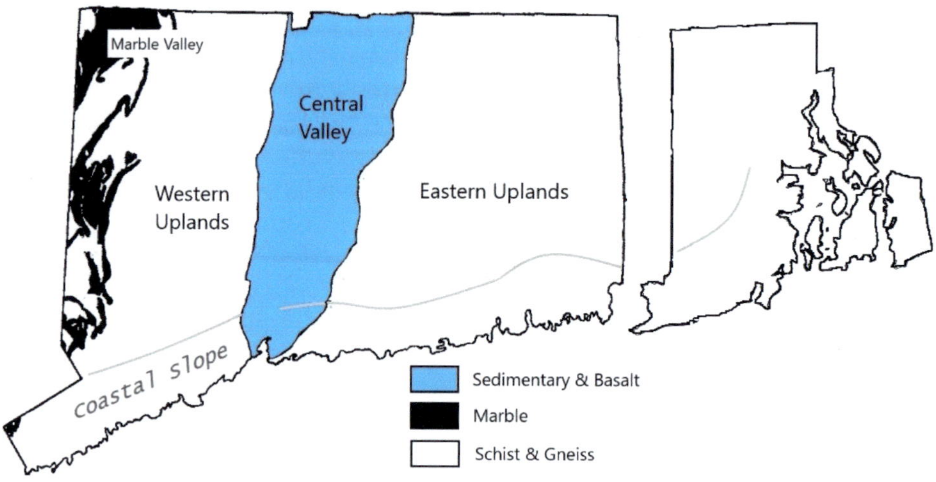

Figure 1. Major geological zones (Adapted from Canavan and Siver 1994).

Features of Aquatic Plants

The ability to distinguish among the many kinds of aquatic plants is based foremost on a knowledge of their forms and characters. Since hydrophytes are derived from terrestrial ancestors, this group of species has developed shared features to overcome the challenges presented with life in water. It is these characteristics that we often rely on in identifying a species.

Keep in mind aquatic plants are notoriously variable in form. Local variations in water depth or flow can influence leaf size, shape, and texture, or presence of flowers or fruits which can make identification difficult—if even possible—at times.

Morphology. True aquatic macrophytes can be submersed (plant anchored to substrate or suspended), floating-leaved (plant anchored to a substrate), or free-floating on water surface (unattached to a substrate). Leaves growing submersed under water are often thin and delicate as to offer little resistance to potentially destructive water actions and to permit more light to penetrate tissues. Likewise, submersed foliage is commonly ribbon-like (fig. 2a) or highly divided (fig. 2c,d) in form. Submersed stems, like leaves, are often thin and supported by water, lacking strengthening agents in their tissues. Consequently, submersed stems and leaves are generally limp and collapse when held out of water. Stems and leaves of certain species are specialized to be buoyant. These organs have air trapped in swollen tissues to reach light closer to the water surface or aid in keeping flowers above water. Floating leaf blades are usually thicker, at times leathery, and capable of shedding water (fig. 2b). Some species possess both floating and submersed leaves on the same plant (fig. 2c).

Reproduction. As their terrestrial relatives do, aquatic plants rely on sexual reproduction via flowers (or spores outside of the angiosperms). In the majority of hydrophytes, flowers are held above water, by various means, to utilize insect pollinators or wind. The insect-pollinated flowers are often showy, colorful (yet white in many species), and sometimes fragrant. In contrast, some hydrophytes develop flowers below water and employ very unusual underwater pollination mechanisms.

Aquatic plants are noted for their ability to reproduce and spread, often aggressively, by asexual means through rapid growth, fragmentation, and or formation of dormant structures. Due to this propensity, hydrophytes are some of our most troublesome weeds (e.g., *Myriophyllum spicatum*, *Hydrilla*). Specialized overwintering, vegetative organs called turions, or "winterbuds", are prevalent among aquatic plants. These dormant, condensed shoots are found mostly in submersed and free-floating species. They are particularly common in species of *Potamogeton*, *Myriophyllum*, and *Utricularia*, and often useful for identification purposes.

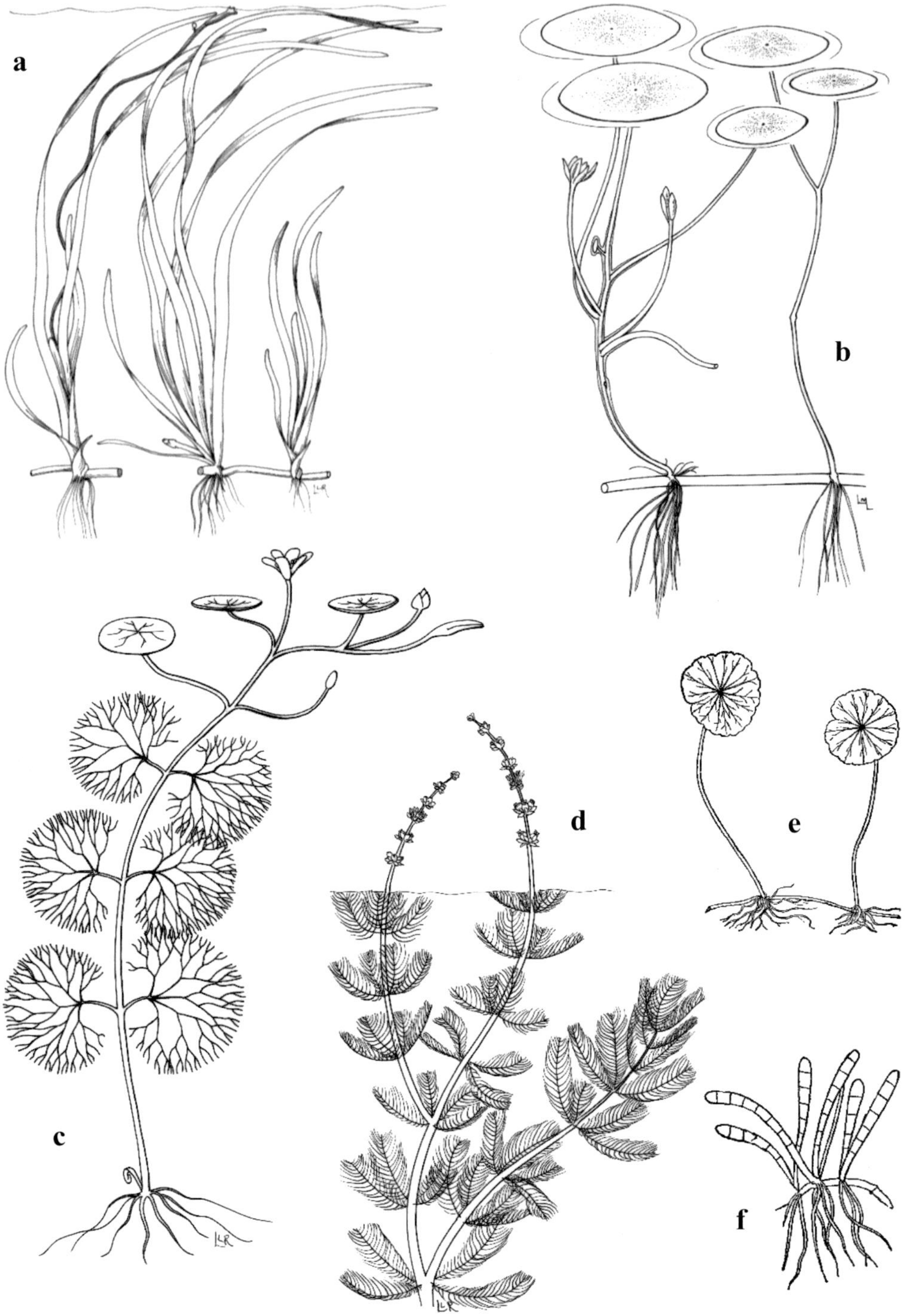

Figure 2. a. *Vallisneria*, with submersed leaves; b. *Brasenia*, with floating leaves; c. *Cabomba*, with submersed and floating leaves; d. *Myriophyllum*, submersed plant with emersed spike; e. *Hydrocotyle*, floating leaves on horizontal stems in substrate; f. *Lilaeopsis*, submersed leaves on horizontal stems in substrate

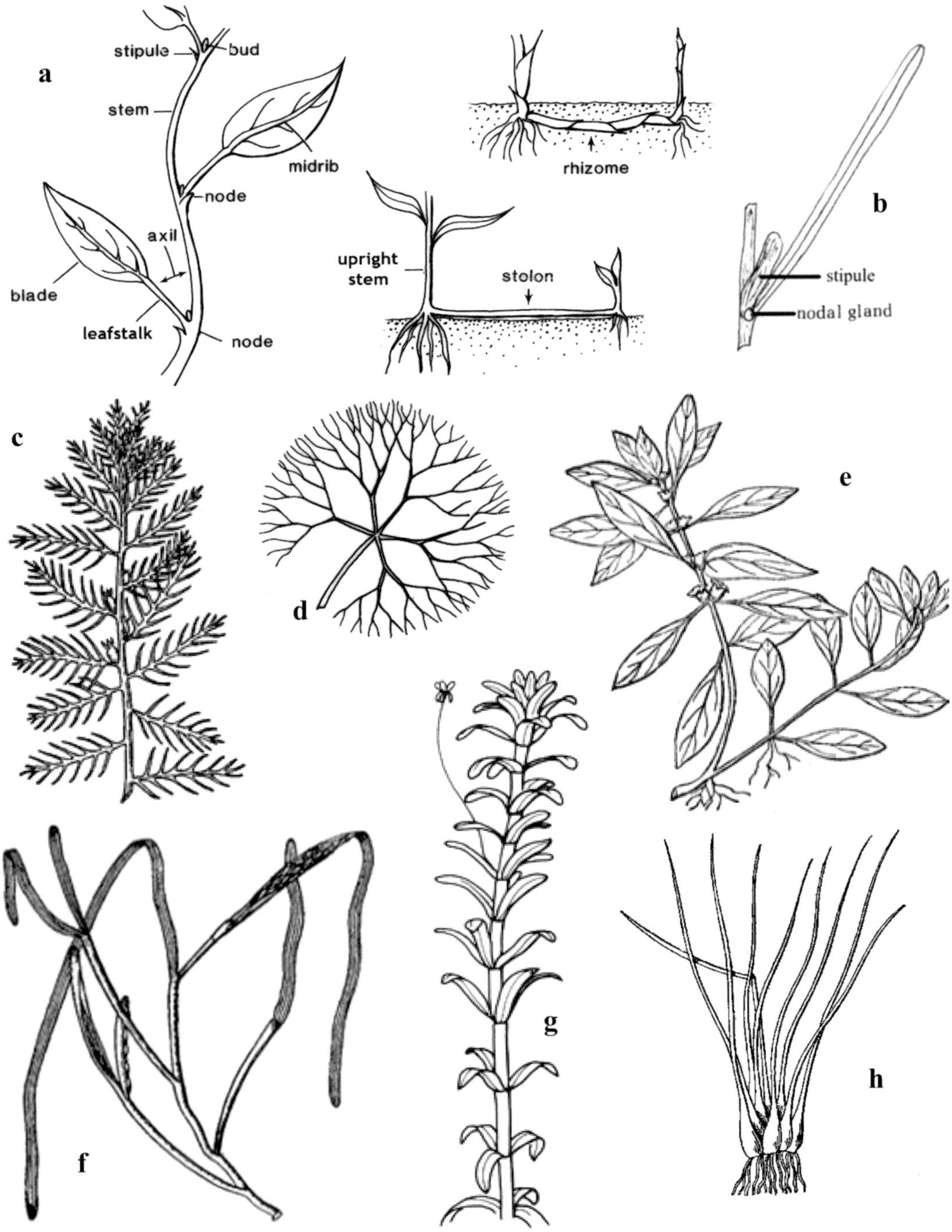

Figure 3. a. Generalized plant morphology; b. *Potamogeton*, node; c. *Proserpinaca*, pinnately divided leaves; d. *Cabomba*, palmately divided leaf; e. *Ludwigia*, undivided, opposite leaves; f. *Zostera*, ribbon-like, alternate leaves; g. *Elodea*, whorled leaves; h. *Isoetes*, basal leaves.

KEYS TO GENERA

SYNOPSIS OF KEYS

1. Plants with stems and leaves entirely submersed, normally growing completely beneath the water's surface (may be temporarily exposed to air with receding water or low tides); only flowering structures may sometimes be held above water surface (fig. 2a,d)…………… **Key I**
1. Plants with stems or leaves exposed to air either by floating on, or held above, water surface; flowering structures may or may not also be held above water surface……………2
   2. Plants with at least some leaves floating on the water surface (fig. 2b,c)……………**Key II**
   2. Plants with some leaves or stem portions extending above the water surface………**Key III**

KEY I: SUBMERSED PLANTS

1. Leaves or leaf-like stems scattered along an upright stem (fig. 2c,d)………………………… 2
1. Leaves or leaf-like stems attached basally (figs. 2a, 3h) or grouped on horizontal stems in substrate (fig. 2f) ……………………………………………………………………………….36
   2. Leaves or leaf-like branches divided into multiple narrow segments (figs. 2d, 3c,d)……… 3
   2. Leaf blade undivided (fig. 3e-g) ……………………………………………………………17
3. Leaves or leaf-like branches whorled along stem (figs. 2d, 3g) ……………………………….4
3. Leaves opposite or alternate along stem (figs. 2c, 3e,f)………………………………………….8
   4. Leaf-like branches bearing small bladders at tips; flowers purple………………..*Utricularia*
   4. Plants lacking bladders at tips; flowers not purple………………………………………. 5
5. Leaf segments serrated with small teeth, leaves somewhat stiff; plants rootless…………… ………………………………………………………………………………..*Ceratophyllum*
5. Leaf segments without teeth, leaves soft in texture; plants rooted…………………………6
   6. Submersed leaves repeatedly divided from base; flowers borne in a conspicuous, yellow, daisy-like cluster………………………………………………………………… *Bidens*
   6. Submersed leaves feather-like, with a central axis; flowers inconspicuous ……………….7
7. Leaves crowded at top of stem; plants often with floating clusters of flowers on swollen, air-filled, spongy stalks in spring/early summer……………………………………. *Hottonia*
7. Leaves scattered along most of stem length; flowers either submersed in leaf axils or emersed on narrow spikes, throughout most of season………………………*Myriophyllum*
   8. Leaves opposite…………………………………………………………………………….9
   8. Leaves alternate……………………………………………………………………………11
9. Submersed leaves pinnately divided, with a central axis; upper portion of stem often with a floating cluster of triangular to diamond-shaped blades and stout, spongy stalks…….. *Trapa*
9. Submersed leaves palmately divided (repeatedly divided from base); lacking a floating cluster of triangular to diamond-shaped blades……………………………… ………10

10. Leaves with a distinct stalk (1–4 cm long); flowers white; a few inconspicuous, narrow (1–4 mm wide) floating leaves also produced at tips of stems during periods of flowering only... *Cabomba*

10. Leaves stalkless (crowded and appearing whorled); flowers yellow, in a daisy-like cluster; emersed, undivided toothed leaves sometimes present.......................... *Bidens*

11. Plants with leathery stems and creeping, green roots adhering to ledge, rocks, or boulders with fleshy disks; plants of rocky, fast-flowing rivers and streams................. *Podostemum*

11. Plants lacking fleshy, adhesive disks; plants of lakes, ponds, or sluggish rivers and streams 12

12. Leaves without one central axis...13

12. Submersed leaves each with a central axis..14

13. Plants with small sac-like bladders, mostly scattered on sides of leaf-like segments; plants rootless, mostly free-floating or anchored by buried branches; flowers yellow...... *Utricularia*

13. Plants lacking bladders; plants rooted to substrate; flowers white or yellow...... *Ranunculus*

14. Plants terminated with a floating rosette (to 50 cm diam.) of broad, triangular to diamond-shaped leaves; fruit large (2–4.5 cm wide), hard, black, and spiny................ *Trapa*

14. Plants without a floating rosette of broad leaves; fruit smaller and without spines.........15

15. Leaves crowded at top of stem; in spring/early summer, plants often with floating clusters of flowers with swollen, spongy, stalks .... *Hottonia*

15. Leaves attached along most of stem length; flowers either submersed or emersed, and without swollen, air-filled, spongy stalks...16

16. Leaves strictly alternate, not crowded along stems; flowers borne in axils of emersed leaves only, each with 3 sepals.. *Proserpinaca*

16. Leaves alternate, or both alternate and whorled on same plant, crowded along stems; flowers borne in axils of submersed leaves or arranged on emersed spikes, each with 4 sepals... *Myriophyllum*

17. Leaves whorled along stem...18

17. Leaves opposite or alternate... 20

18. Leaves in whorls of 4–8...19

18. Leaves in whorls of 3 (some leaves opposite in pairs) .... *Elodea*

19. Leaves mostly in whorls of 4, blades 15–40 mm long, margins lacking conspicuous teeth..... .... *Egeria*

19. Leaves in whorls of 4–8, blades 8–15 mm long, margins conspicuously saw-toothed.......... .... *Hydrilla*

20. Leaves opposite..21

20. Leaves alternate...28

21. Submersed leaves linear 22

21. Submersed leaves lance-shaped, egg-shaped, or rounded...24

22. Leaf margins with minute teeth (visible with hand lens), blades abruptly widened at base and sheathing stem... *Najas*

22. Leaf margins without teeth; blades not widened at base 23

23. Submersed leaves (>1 mm wide) notched at tip; stipules lacking; plants with or without a cluster of floating egg-shaped leaves near stem tip; fruit 2-lobed............................ *Callitriche*

23. Submersed leaves (≤1 mm wide) pointed at tips; stem sheathing stipules at base of leaves; plants without floating leaves; fruit not lobed, slightly toothed on outer edge..... *Zannichellia*

24. Plants diminutive (usually <5 cm tall) and mat-forming..25

24. Plants larger (usually >5 cm tall) and not mat-forming..27

25. Plants somewhat fleshy and stout; leaves lance-shaped and pointed at tips........................26

25. Plants not fleshy; leaves lance-shaped, egg-shaped, or rounded, and rounded, pointed, or

notched at tips.. *Elatine*

26. Bases of opposite leaves fused to form a boat-shaped cup at each node; flowers borne singly in leaf axil.. *Crassula*

26. Bases of leaves not forming a boat-shaped cup; plants sterile.................... *Gratiola*

27. Leaves small (6 mm long), rounded, and without stalks; stems flaccid and mostly unbranched; plants without flowers.. *Hypericum*

27. Leaves larger (10–50 mm long), oval to egg-shaped, and conspicuously stalked; stems many-branched and reddish; flowers borne singly in leaf axils........................ *Ludwigia*

28. Leaves inconspicuous and reduced to minute, stubby scales (<1 mm long); stems mostly unbranched arising singly from buried rhizomes................................ *Myriophyllum*

28. Leaves conspicuous (>3 cm long); stems branched..29

29. Submersed leaves linear, ribbon-like or thread-like, 0.2–12 mm wide, with or without an evident mid-vein, blades lacking a stalk; sheathing stipules present or absent at leaf base... 30

29. Submersed leaves elliptic, oblong, or lance-shaped, 3–75 mm wide, with an evident mid-vein, blades with or without a stalk; sheathing stipules present at leaf base...... *Potamogeton*

30. Submersed leaves flat and ribbon-like, 1–12 mm wide; stipules may or may not be present at leaf base..31

30. Submersed leaves thread-like <1 mm wide; stipules present at leaf base................34

31. Leaves with an evident mid-vein; stipules present (2–100 mm long); plants may also have stalked floating blades; flowers and fruits borne in cylindrical spikes................ *Potamogeton*

31. Leaves without a mid-vein; stipules lacking; floating blades absent; flowers and fruits otherwise..32

32. Leaves to 10 cm long, 1–5 mm wide; flowers yellow with six petal-like lobes, borne singly; fruit an elongated capsule.. *Heteranthera*

32. Leaves 30–100+ cm long, 2–12 mm wide; flowers not yellow, borne in clusters; fruit a beaked drupe..33

33. Leaves not sheathing the stem; flowers borne at or above water surface in dense, spherical clusters; plants of freshwaters .... *Sparganium*

33. Leaves sheathing the stem; flowers submersed, enclosed within a leaf-like sheath; plants strictly marine.. *Zostera*

34. Leaves 1-veined and blunt-tipped; floating blades absent; flowers and fruits borne underwater; fruits in clusters of four, each fruit on a long stalk (1–2 cm long); plants chiefly coastal in saltwater, with thread-like rhizomes.................................... *Ruppia*

34. Submersed leaves with 1–5 veins and sharp or blunt tips; floating blades sometimes present; flowers and fruits borne on crowded, cylindrical spikes mostly above water; plants of freshwaters, mostly lacking rhizomes..35

35. Leaves stiff, channeled, 1-veined, and sharp-tipped; floating blades absent; rhizomes bearing small (1–1.5 cm), white tubers.. *Stuckenia*

35. Leaves flaccid, not channeled, with 1–5 veins, and sharp or blunt tips; plants sometimes with floating blades; rhizomes mostly lacking (if present, not bearing tubers)................ .... *Potamogeton*

36. Leaves or leaf-like stems thread-like or quill-like (linear and rounded in cross-section) 37

36. Leaves flattened and/or more ribbon-like..43

37. Leaves clustered into rosettes.. 38

37. Leaves or leaf-like stems attached singly or grouped into clusters along horizontal stems 40

38. Roots with cross walls; flowers, if present, to 15 mm wide with 3 white petals, and borne in clusters.. *Sagittaria*

38. Roots without cross walls; flowers, if present, small (3 mm wide) with 5 white petals, and borne singly..39

39. Horizontal stems absent; leaves hollow, with a few cross-walls, swollen and whitish at bases, mostly >5 cm long; flowers/fruits always absent................................... *Isoetes*

39. Horizontal stems present, thread-like; leaves fleshy, narrow at base, ≤5 cm long; flowers/fruits may be present.. *Limosella*

40. Leaves fleshy, 1–2 mm wide; flowers white... *Limosella*

40. Leaves or leaf-like stems thread-like (<0.5 mm wide) and delicate; flowers yellow, purple, or brown..41

41. Horizontal stems thick (3–4 cm diameter)...*Juncus*

41. Horizontal stems, if present, thread-like (<2 mm diameter).................................42

42. Flowers showy, yellow or purple; leaves completely or mostly hidden in substrate; plants only apparent when flowering; a few tiny (0.2–1.1 mm long) sac-like bladders may be present, also buried in substrate...*Utricularia*

42. Flowers not showy, borne in small (2–8 mm long) clusters behind overlapping brown scales; leaf-like stems upright, not hidden in substrate; plants apparent with or without flowers; bladders not produced...*Eleocharis*

43. Leaves thin, flaccid, and ribbon-like...44

43. Leaves flattened but thicker, not ribbon-like and flaccid.................................46

44. Leaf blade with a 3-zone appearance (light-colored middle zone bordered by darker zone on sides); flowers small (2 mm wide) floating at surface with transparent petals; fruit a cylindrical capsule borne at end of a long, corkscrew-like stalk......................... *Vallisneria*

44. Leaf blade without a 3-zone appearance; plants often sterile, but, if flowers present, larger and emersed above water; if fruit present, not borne on a long, corkscrew-like stalk..........45

45. Roots with cross walls; leaves with cross veins running perpendicular to central mid-vein, blades often long (>25 cm).. *Sagittaria*

45. Roots without cross walls; leaves without cross veins, blades 6–25 cm long.....*Pontederia*

46. Plants low and small, leaves mostly 1–6 cm long...47

46. Plants larger, leaves to 35 cm long, sometimes thick and spongy......................... *Sagittaria*

47. Leaves hollow...48

47. Leaves not hollow...49

48. Leaves numerous, each with two hollow chambers in cross section; flowers conspicuous, clustered on long (>20 cm long) stalks; fruit a nodding capsule........................... *Lobelia*

48. Leaves 1–few, each hollow but transversely divided with 4–6 cross walls; flowers inconspicuous, and borne on stalks about as long as the leaves; fruit a nutlet with 5 corky ribs.. *Lilaeopsis*

49. Leaves attached in pairs along slender rhizomes, round and widest at tips; flowers borne singly on very short (0–2 mm) stalks between the leaf pairs..................*Glossostigma*

49. Leaves numerous in tufted rosettes, tapered at tips; flowers borne in clusters on longer stalks (>2.5 cm)..50

50. Roots with conspicuous cross walls; leaves widest at base and with conspicuous cross-veins; spherical whitish flower clusters borne at end of a vertically ridged stem... .*Eriocaulon*

50. Roots without cross walls; leaves lance-linear in shape, without cross-veins; a few white-petaled flowers borne at end of non-ridged stem.. *Helanthium*

Key II: Floating & floating-leaved plants

1. Whole plant is free-floating or suspended at water surface; plants either lacking roots or with roots extending into water column...2
1. Only the leaves or branches are normally floating at water surface; plants typically rooted in the substrate (fig. 2b, e)..8
   2. Plants diminutive, ≤1.5 cm long; leaves not in rosettes; roots, if present, unbranched........5
   2. Plants larger; leaves 2–40 cm long, arranged in rosettes; roots branched and feathery......3
3. Leaves without stalks, light green, and velvety-hairy.......................................*Pistia*
3. Leaves with obvious stalks, dark green, and glossy...4
   4. Leaves with greatly swollen/inflated stalks, blades not spongy; flowers large (5–7 cm wide), showy, and blue or purplish-blue; roots black..................................... *Eichhornia*
   4. Leaves with stalks not inflated, blades spongy; flowers not showy, whitish; roots not black...*Limnobium*
5. Leaves distinguished from stems; leaf blades fuzzy and positioned in two rows along horizontal stems...*Azolla*
5. Plant body not differentiated into distinct stems and leaves; plants not fuzzy.................6
   6. Plants with roots below, extending into water; bodies flattened throughout, 1–4 mm wide...7
   6. Plants without roots; bodies spherical, oval, or boat-shaped (i.e., flattened above and rounded below), 0.5–1 mm wide.. *Wolffia*
7. Plants each with one root below; bodies green beneath.....................................*Lemna*
7. Plants each with 7–21 roots below; bodies reddish purple beneath........................ *Spirodela*
   8. Plant with a floating rosette of noticeably inflated (swollen and spongy) leaf stalks or branches; submersed leaves finely divided with thread-like segments........................... 9
   8. Plant lacking a floating rosette of inflated branches or leaf stalks; submersed leaves, if present, not finely divided...11
9. Floating leaves or upright branches green; bladders absent on plant; flowers white..........10
9. Wheel-like rosette of whitish branches floating flat on water surface, functioning as a float for the flowering stem; numerous bladders present on submersed portion of plant; flowers yellow... *Utricularia*
   10. Rosettes of floating large-toothed, triangular to diamond-shaped, flat blades; flowers conspicuous; fruit large (20–45 mm wide) with stout spines......................................*Trapa*
   10. Rosettes of erect stalks, lacking flat leaf blades, instead bearing whorls of inconspicuous flowers; fruit small (1.5–3 mm wide) lacking spines.......................................*Hottonia*
11. Floating leaves linear and ribbon-like, > 10 times as long as wide, without obvious stalks, usually just tips of leaves floating..12
11. Floating leaves with broad blades, not ribbon-like, with obvious leaf stalks; usually the entire blade is floating..15
   12. Leaves 10–15 mm wide, with conspicuous cross veins (short veins running perpendicular to longer veins); flowers borne either below water or on floating stems.........................13
   12. Leaves 2–10 mm wide, without cross veins; flowers borne above water on emergent stems...14
13. Leaf with a 3-zone appearance (a light-colored middle zone bordered by a darker zone on each side), blade not widening at tip; flowers with transparent petals; fruit cylindrical (5–12 cm long), borne singly at end of a coiled stalk; roots without cross walls................*Vallisneria*
13. Leaf without a 3-zone appearance, blade sometimes widening at tip; flowers showy, each with three white petals, borne in clusters at end of a long, usually floating, stem; fruit

flattened and seed-like (to 2.5 mm long), crowded into spherical clusters; roots with cross walls.. *Sagittaria*

14. Leaves 15–45 cm long, dark bluish-green, with a prominent central vein; base of leaf forming a tube-like sheath around the stem; flowers and fruits in minute, narrow (1–2.5 mm wide), soft clusters..*Glyceria*

14. Leaves 50–150 cm long, green, lacking a central vein; base of leaf not tubular; flowers in spherical (5–25 mm wide) clusters developing into hard, bur-like fruit clusters...*Sparganium*

15. Leaf stalks centrally attached to underside of floating leaf blade..............................16

15. Leaf stalks attached at the edge of floating leaf blade.. 18

16. Leaf blades are broadly oval; submerged portions of plant covered in a clear, jelly-like coating; stems flexible and buoyant; flowers are red purple, borne singly.............*Brasenia*

16. Leaf blades are circular; submerged portions of plant not covered in a jelly-like coating; stems creeping on substrate; flowers are whitish or yellow, borne singly or in clusters.....17

17. Plants typically massive with large (to 60 cm wide), blue green, leaf blades attached to long (to 2 m) stout stalks; flowers large (10–20 cm wide), yellow or pink, and borne singly.... .... *Nelumbo*

17. Plants with smaller (1–7.5 cm wide), green, leaf blades attached to shorter (to 25 cm long) stalks; flowers minute, whitish, and borne in clusters of 15–35........................*Hydrocotyle*

18. All leaves are basal, usually arising from base of plant close to substrate.....................19

18. Most leaves attached along an upright or horizontal stem in the water column..........25

19. Floating leaves divided into segments or deeply lobed along sides..............................20

19. Floating leaves not divided into segments or lobed along sides (blades may have a basal sinus)...21

20. Floating leaf comprised of 4 triangular segments (each 7–21 mm long, 6–19 mm wide), resembling a four-leaf clover; reproductive structures (not flowers) submersed, hairy, pea-shaped (5 mm), near base of leaf stalks..*Marsilea*

20. Floating leaf simple, blade deeply lobed and divided into 3–5 main segments; flowers emersed above water on branching clusters, each with five, small (2–5 mm long), yellow petals... *Ranunculus*

21. Floating blades blue green, parallel-veined, oblong to elliptical, without a sinus; flowers minute, without petals, and borne in dense clusters at end of thick, white stalks.....*Orontium*

21. Floating blades green to purplish, veins centrally radiating and/or branching, generally round or heart-shaped, with a deep sinus; flowers conspicuous, with petals, not borne on thick, white stalks...22

22. Plants also with linear, submersed, phyllodial leaves (25–450 mm long, 15–110 mm wide); roots with cross walls; fruits flattened and seed-like (to 2.5 mm wide), crowded into spherical clusters...*Sagittaria*

22. Plants lacking linear, submersed leaves; roots without cross walls; fruits globose, berry-like (4–40 mm wide) and borne singly...23

23. Leaves large (mostly >8 cm wide), veins branched and running toward margins, lacking a spongy pad on blade underside; flowers bisexual, showy, yellow, white, or pinkish..........24

23. Leaves smaller (up to 8 cm wide), parallel-veined, veins running from base and converging toward tip, with conspicuous spongy pad on blade underside...................... *Limnobium*

24. Leaf blade generally circular in shape, with a closed sinus and pointed lobes, and usually purplish on underside; leaf stalk round in cross section; flowers 5–19 cm wide with numerous, white to pink, petals; fruits maturing underwater on coiled stalks......*Nymphaea*

24. Leaf blade generally heart-shaped, with an open sinus and rounded lobes, and green on underside; leaf stalk flattened; flowers 1–4.5 cm wide with 5–6 bright yellow, petal-like sepals; fruits maturing at water surface on straight stalks..............................*Nuphar*

25. Floating leaves generally round or heart-shaped; flowers are emersed, showy, 5–25 mm wide, yellow or white..26
25. Floating leaves elliptical, oblong, or egg-shaped; flowers are small (<5 mm wide), submersed or emersed, never yellow or white...27
26. Floating leaves with a deep sinus at base, veins centrally radiating and branching, purplish on undersides; submersed clusters of slender fleshy roots may be present at the leaf base; flowers with 5 yellow or white petals and 5 stamens........................ *Nymphoides*
26. Floating leaves shallowly lobed at base, with parallel veins, green on undersides; lacking clusters of fleshy roots at the leaf base; flowers with 6 white petals and 3 stamens.. *Heteranthera*
27. Floating leaves tightly clustered into a rosette at the stem tip, blades small (5–30 mm long) and widest toward the tip; submersed leaves arranged in opposite pairs and linear; stipule absent at base of each leaf; flowers borne 1–3 in submersed leaf axils........*Callitriche*
27. Floating leaves not clustered into a rosette, blades usually much larger (6–260 mm long) and widest near middle or base of blade; submersed leaves, if present, linear or broad, and alternate; stipule usually present at base of each leaf; flowers borne on emersed cylindrical spikes or submersed few-flowered spike-like clusters..28
28. Leaf veins parallel; floating and submersed leaves different in shape and texture; leaves never hairy; flowers greenish..*Potamogeton*
28. Leaf veins pinnately branching; floating and submersed leaves no different; leaves sometimes hairy; flowers pink..*Persicaria*

Key III: Plants with some leaves or stem portions extending above the water surface

1. Leaves or leaf-like branches held above water are linear, >10 times as long as wide............ 2
1. Leaves held above water with a more expanded blade, <10 times as long as wide.............. 9
2. Stems held above water swollen and clustered into a floating rosette, spongy stalks (to 30 cm long), bearing whorls of tiny white flowers; submersed leaves are feathery, finely divided into linear segments..*Hottonia*
2. Leaves or stems above water not in floating rosettes; submersed leaves not feathery..... 3
3. Plants with many submerged, flaccid, thread-like leaves or stems...................................4
3. Plants without submerged, flaccid, thread-like leaves or stems...6
4. Plants with 1 (or maybe 2) fully developed, emergent leaves attached to the main stem, leaf with internal cross walls; submerged thread-like leaves with internal cross walls; flower cluster multi-branched...*Juncus*
4. Emergent leaves lacking (no leaves attached to main stem); submerged thread-like leaves or stems lacking internal cross walls; flower cluster unbranched.....................................5
5. Erect, emergent stems round in cross section, narrower than flower cluster; flower cluster appears to come from side of stem (because a 2–20 mm long leaf-like bract is attached below flower cluster, appearing like a continuation of the stem); submerged leaves with a sheathing base..*Schoenoplectus*
5. Erect, emergent stems 3-sided in cross section, nearly as thick as the flower cluster; flower cluster attached to tip of stem; submerged leaf-like stems without a sheathing base.. *Eleocharis*

6. Leaves clustered in a basal rosette, each essentially round in cross-section and thickest at base, generally tapering to the tip; flowers, if present, with petals.......................... 7
6. Leaves not in a basal rosette, each generally flattened or thread-like; flowers are lacking petals 8
7. Base of plant swollen and bub-like, containing white spores; roots are not septate; leaves hollow, flexible; flowers, fruits or flowering stems never produced...................*Isoetes*
7. Base of plant not swollen and bub-like; roots are septate; leaves rigid; flowers with three conspicuous white petals..*Sagittaria*
8. Leaves conspicuous (>0.5 m long, 2–10 mm wide); flowers are grouped into several spherical clusters, each with a leaf attached below; fruits are a spherical aggregate of hard, beaked drupes.. *Sparganium*
8. Plants delicate, grass-like, with short (5–8 cm tall) and slender (<0.5 mm wide) leaf-like stems; actual leaves scale-like and inconspicuous at stem base; flowers are grouped at the stem tip into a single cluster of overlapping scales; fruits are nut-like with bristles at the base... *Eleocharis*
9. Leaves held above water are deeply divided or finely divided into narrow segments......10
9. Leaves held above water are broad and flattened, not finely divided, but may have toothed margins..12
10. Emersed leaves greatly reduced and bract-like, <1 cm long.........................*Myriophyllum*
10. Emersed leaves well-developed, 1–3.5 cm long, essentially like submersed leaves...........11
11. Leaves whorled, each divided into 10–15 pairs of narrow segments.............*Myriophyllum*
11. Leaves alternate, each divided into 6–9 pairs of narrow segments.................*Proserpinaca*
12. Emersed leaf margins are toothed...13
12. Emersed leaf margins are without teeth..17
13. Submersed feather-like leaves present, attached to upright stems..........................14
13. Submersed feather-like leaves absent, leaves with round blades, attached to creeping stem.. *Hydrocotyle*
14. Leaves held at or above water are attached alternately at top of a submersed stem, each with a triangular to diamond-shaped blade, and a long leaf stalk with an inflated portion; fruit with 4 horn-like spikes... *Trapa*
14. Leaves held above water are alternate, opposite, or whorled on emersed portion of stems, each with a blade much not longer than wide, and a leaf stalk lacking an inflated portion; fruit with without 4 horn-like spikes...15
15. Emersed leaves are opposite or whorled on a stem.. 16
15. Emersed leaves are alternate on stem... *Proserpinaca*
16. Emersed leaves are opposite, 2 per node, and 2–4 cm long; flowers borne in a single, conspicuous, yellow, daisy-like cluster.. *Bidens*
16. Emersed leaves are whorled, 4–6 per node, and 0.4–1.8 cm long; flowers borne in axils of leaves forming a leafy spike.. *Myriophyllum*
17. Emersed leaves opposite, 2 per node, blade 0.5–4 cm long, stalkless or short-stalked (<3 cm long)...18
17. Emersed leaves arising from a basal rosette, blade 3–30 cm long, on long stalks (>6 cm long)..19
18. Emersed leaves without stalks; stems green; flowers borne in a single, conspicuous, yellow, daisy-like cluster at stem tip; submersed leaves are finely divided into thread-like segments...*Bidens*
18. Emersed leaves with stalks (5–25 mm long); stems reddish; flowers are inconspicuous, green-brown, borne singly in leaf axils; submersed leaves resembling emersed leaves.........
.... *Ludwigia*

19. Leaves blue green, leaf stalk somewhat fleshy, about same length as blade…….*Orontium*
19. Leaves green, leaf stalk spongy, its length much greater than that of blade…..…….….20
20. Leaf blades with round basal lobes; fruits are round, berry-like………………….*Limnobium*
20. Leaf blades without lobes or basal lobes pointed; fruits are flattened, winged, and seed-like………………………………………………………………………………….…….*Sagittaria*

Keys to Species & Descriptions

Azolla — Azollaceae
Mosquito fern

Azolla caroliniana Willd. EASTERN MOSQUITO FERN

Eastern Mosquito fern is a tiny, free-floating, aquatic fern with horizontal stems (0.5–1 cm long), slender roots, and overlapping, fuzzy leaves. Leaf blades are floating, very small (0.5–2 mm wide), dark green (or reddish), unequally two-lobed, covered in hairs, and positioned in two rows along a stem. Plants reproduce by spores.

HABITAT & DISTRIBUTION

Azolla caroliniana is a non-native, moss-like, mat-forming fern species of stagnant lakes, ponds, or sluggish streams. Introduced and infrequent in CT.

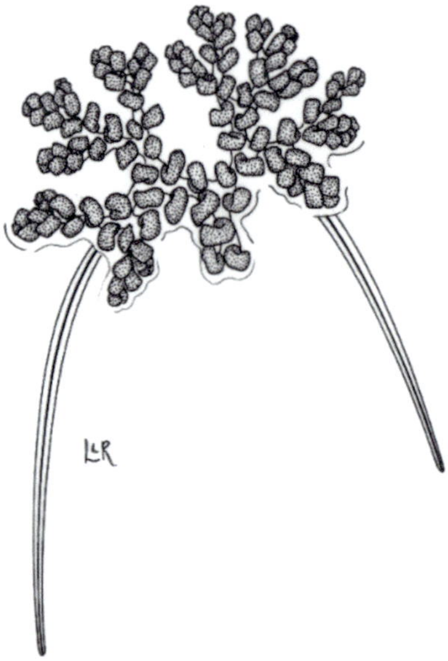

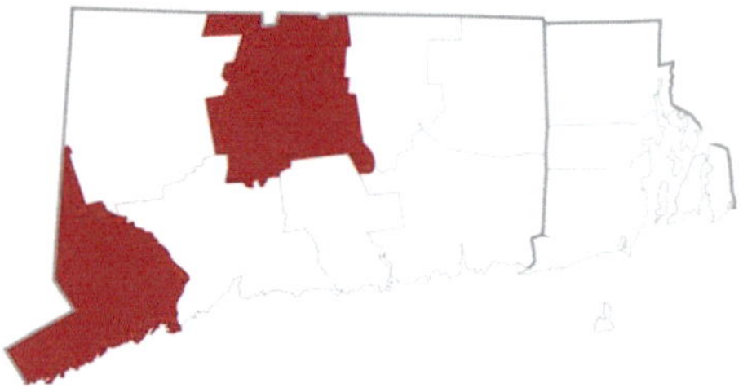

Bidens — Asteraceae
Beggar-ticks

Bidens beckii Torr. ex Spreng. BECK'S BEGGAR-TICKS

Beck's beggar-ticks, or Water marigold, is a submersed, rooted, perennial with rhizomes and elongated (to 200 cm), slender, flexible stems. Leaves are opposite, stalkless, 2–4 cm long, and dimorphic: submersed leaves, appearing whorled, are finely divided into thread-like segments; and emersed leaves are simple, lance-shaped, with toothed or untoothed margins. Flowers are borne, above water, in a single, conspicuous, yellow, daisy-like cluster (head) resembling a flower (2.5 cm diam.). Fruit is a yellowish to greenish brown, linear achene (10–15 mm) with 3–6 barbed extensions (13–15 mm long) at tip.

HABITAT & DISTRIBUTION

Bidens beckii is a native species in lakes, ponds, and slow streams. It is uncommon in CT and RI and state-listed as ***special concern*** in CT. Fruits rarely mature in our region. It can occupy deep (to 7 m) waters where it will only produce submersed foliage.

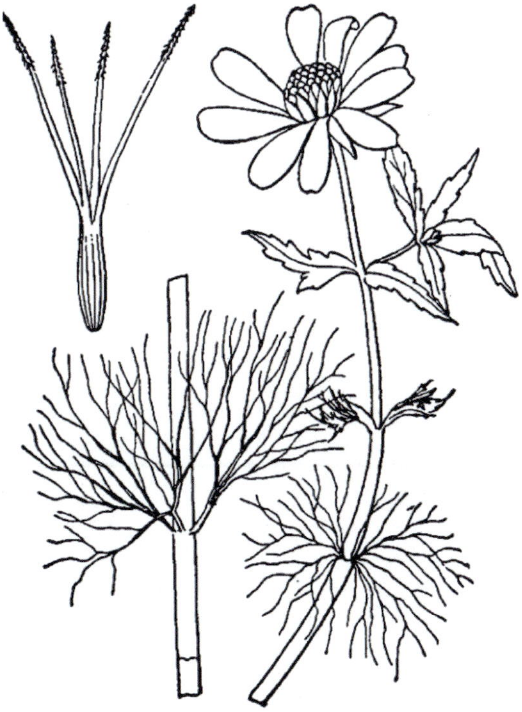

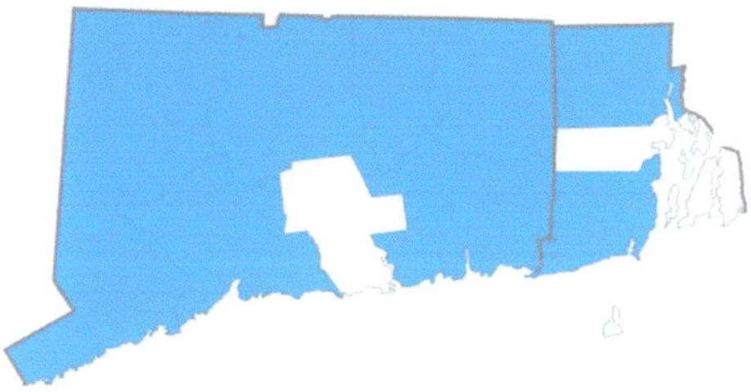

Brasenia — Cabombaceae
Watershield

Brasenia schreberi J.F. Gmel. — WATERSHIELD

Watershield is a rooted, perennial, floating-leaved plant with creeping rhizomes and long, flexible, buoyant, branching stems. Leaves are alternate, floating, broadly oval (4–13 cm long), on long (to 60 cm) stalks centrally attached to the blades. Usually, all submerged portions are purplish and covered in a thick, translucent, jelly-like coating. Flowers (2 cm wide) are bisexual, mostly wind-pollinated, petals red purple (12–20 mm long), borne singly on elongated (to 15 cm) stalks that hold flowers about 2 cm above water. Fruit a small (6–10 mm long), leathery, few-seeded nutlet.

HABITAT & DISTRIBUTION

Brasenia schreberi is native and found in open, standing or slow waters (1–3 m deep) of lakes, ponds, and streams. It is widespread and common in CT and RI. This species is highly tolerant and can reproduce asexually via rhizomes and winter buds which can result in dense, weedy colonies.

Notes: This is a very distinctive floating-leaved plant with its oval leaves and slippery, mucilaginous layer covering younger, underwater parts.

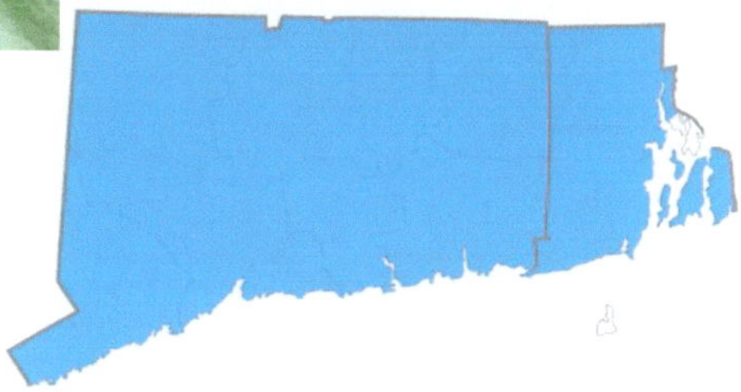

Cabomba — Cabombaceae
Fanwort

CAROLINA FANWORT *Cabomba caroliniana* Gray

Carolina fanwort is a rooted, submersed species with creeping rhizomes and long (up to 2 m), slender, branching, flexuous stems. Submersed leaves are opposite, rotund (1–3.5 cm long, 2.5–5 cm wide), palmately divided (3–7 divisions) into many narrow segments, and with distinct stalk (1–4 cm long). During periods of flowering only, a few inconspicuous, narrow (5–20 mm long, 1–4 mm wide), floating leaves are also produced at tips of stems. Flowers are bisexual, small (6–15 mm diam.), white, and borne singly above water on stalks (3–10 cm) from upper leaf axils. Fruit is a small (4–7 mm long), leathery, nutlet.

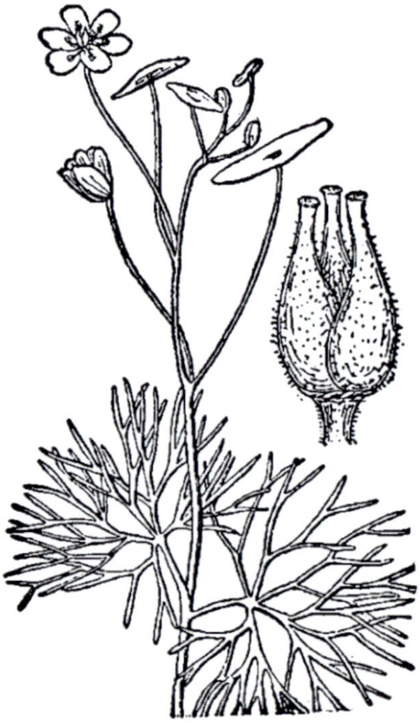

HABITAT & DISTRIBUTION

Cabomba caroliniana is non-native and has become naturalized in ponds and slow streams. It is common in CT and RI where it is **invasive** and prohibited. This plant can become abundant and problematic as it exhibits aggressive vegetative growth, prolific stem fragmentation, and broad ecological tolerances.

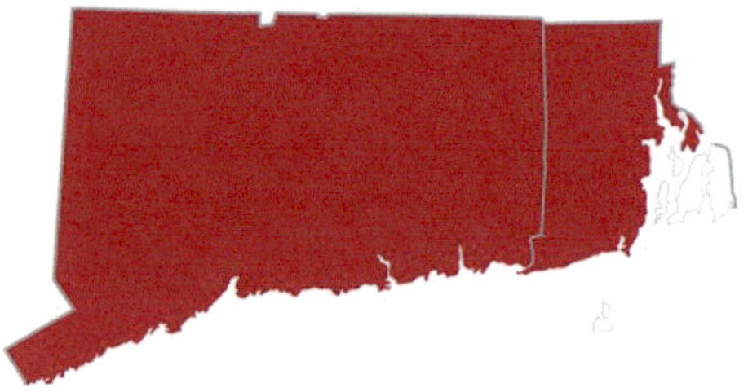

Callitriche — Plantaginaceae

Water starwort

The Water starworts are small, submerged, rooted, annuals or perennials, with slender, branched stems. Leaves are opposite, toothless, and mostly dimorphic: submersed leaves are linear shaped, often notched at tip, and well-spaced along the stem, and floating, broader leaves are clustered into a rosette near the stem apex. In most species (except *C. hermaphroditica*), the bases of adjacent leaves are slightly confluent, forming a narrow membranous wing across the sides of the stem (This feature is not always easily detected). Flowers are tiny, unisexual (on same plant), petal-less, and borne 1–3 per leaf axil. Female flowers have a flattened, 2-lobed ovary. Fruit splitting into four 1-seeded segments at maturity.

Four species occur in CT and RI, including one non-native. The species are very variable vegetatively. Mature fruits are essential for accurate identification.

KEY TO SPECIES

1. Flowers and fruits with 2 translucent bracts (0.5–1.5 mm long) at base; leaves along the stem linear, while those crowded near the stem tip spoon-shaped, forming a floating rosette……………………………………………………………………………………2
1. Flowers and fruits without bracts at base; leaves all linear, never forming a floating rosette.. ……………………………………………………………………………*C. hermaphroditica*
    2. Fruit 1.5–2 mm long, margins with a prominent wing extending along entire fruit length; floating leaf blades 5–8 mm wide with 5–7 veins………………………… *C. stagnalis*
    2. Fruit 1–1.4 mm long, margins unwinged or with a wing absent at fruit base; floating leaf blades (when present) up to 5 mm wide, with 3 (–5) veins…………………………………3
3. Fruit longer than wide by 0.2 mm or more, top portion of each segment with a thin wing, pits of fruit surface aligned in vertical rows……………………………………… *C. palustris*
3. Fruit as wide as long, lacking wings along margins, pits on fruit surface not aligned in rows ……………………………………………………………………………… *C. heterophylla*

C. heterophyllum

C. palustris

C. stagnalis

AUTUMN WATER STARWORT *Callitriche hermaphroditica* L.

Autumn Water starwort is a wholly submersed plant with loosely clustered stems (5–15 cm long). Leaves are all submersed, linear (4–20 mm long, 0.6–2.2 mm wide), widest at base, 1-veined, and notched at tip, and usually widely spaced along stem. Fruits rounded (1–2 mm long) with subtle but discrete winged margins.

HABITAT & DISTRIBUTION

Callitriche hermaphroditica is a native species of lakes, ponds, and rivers. It is uncommon in CT. Plants are generally found in soft, nutrient-enriched waters.

Notes: Leaves tend to be darker green compared to submersed leaves of other *Callitriche* species.

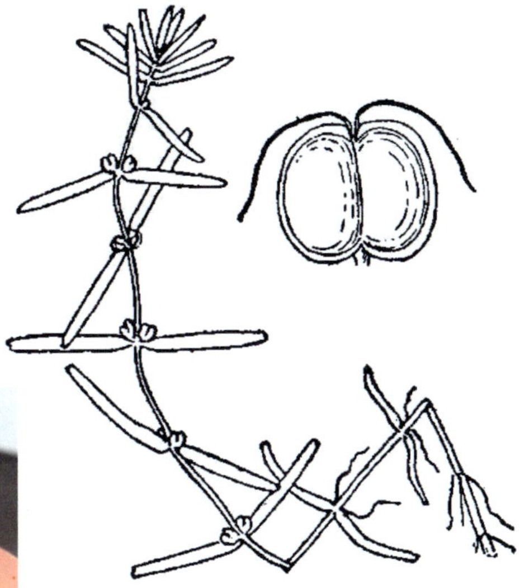

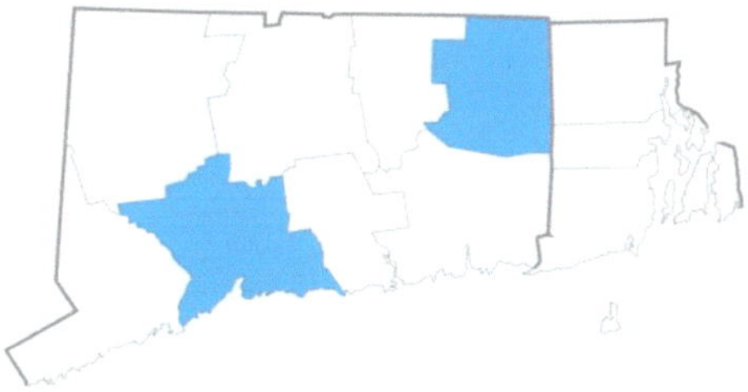

GREATER WATER STARWORT *Callitriche heterophylla* Pursh

Greater Water starwort, or Two-headed Water starwort, is a submersed plant with loosely clustered stems (10–20 cm long). Floating leaf blades (when present) are egg-shaped (5–30 mm long, 2.5–5 mm wide), widest toward the apex, blunt tipped, usually with 3 veins. Submersed leaves are linear (5–25 mm long, 1.5 mm wide) and notched at tip. Fruits about as long as wide (1–1.4 mm long, 0.6–1.2 mm wide) with unwinged margins.

HABITAT & DISTRIBUTION

Callitriche heterophylla is a native species of lakes, ponds, bogs, vernal pools, and rivers. It is widespread in CT and RI. Plants are generally found in near neutral, shallow waters.

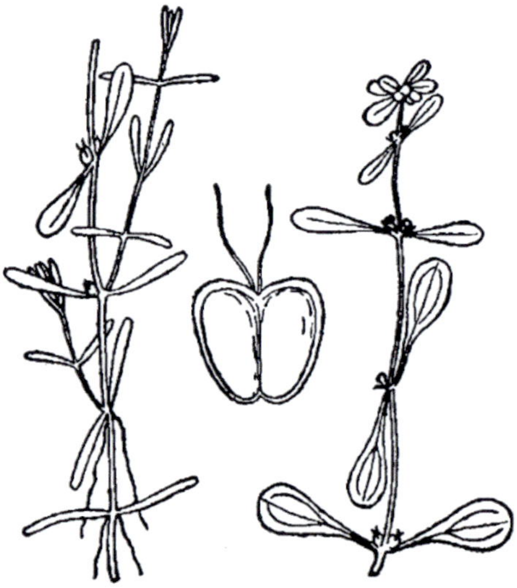

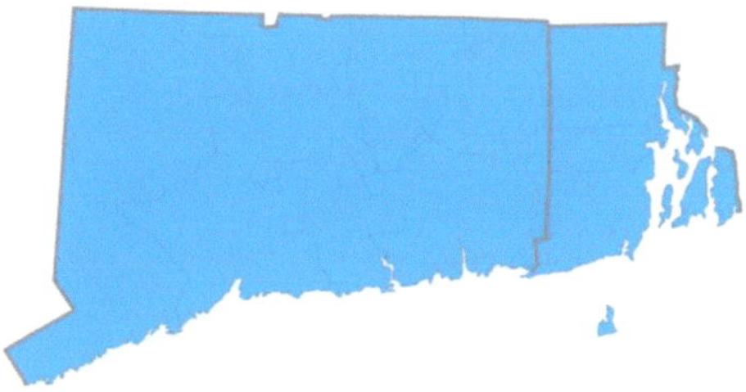

VERNAL WATER STARWORT *Callitriche palustris* L.

Vernal Water starwort is a submersed plant with loosely clustered stems (10–20 cm long). Floating leaf blades (when present) are egg-shaped (5–30 mm long, 2.5–5 mm wide), widest toward the apex, blunt-tipped, usually with 3 veins. Submersed leaves are linear (5–15 mm long, 3 mm wide) and notched at tip. Fruits distinctly grooved, slightly longer than wide (1–1.4 mm long), widest above middle, with a thin wing along upper portions of each lobe.

HABITAT & DISTRIBUTION

Callitriche palustris is a native species of springs, marshes, bogs, fens, ponds, and slow rivers. It is widespread in CT and RI. **Note:** This species can grow as an emergent plant when exposed on wet shorelines or depressions.

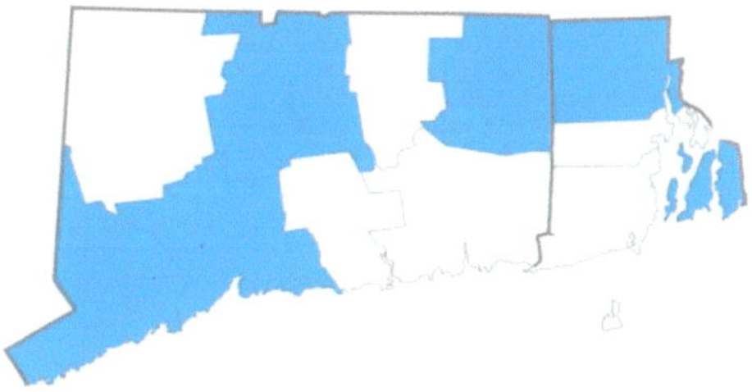

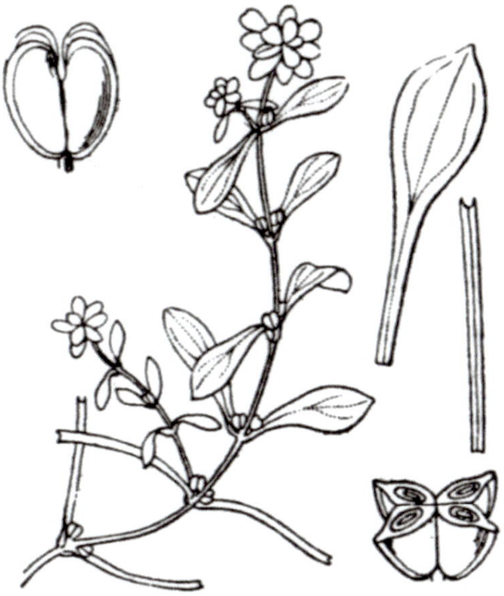

Pond Water starwort *Callitriche stagnalis* Scop.

Pond Water starwort is a submersed plant with loosely clustered stems (10–20 cm long). Floating leaf blades are egg-shaped (to 20 mm long, 5–8 mm wide), widest toward the apex, blunt-tipped, usually with 5–7 veins. Submersed leaves are usually linear (4–10 mm long), 1-veined, and notched at tip. Fruits distinctly grooved, nearly round (1.5–2 mm long), w ith a prominent wing extending evenly along whole length of margin.

Habitat & Distribution

Callitriche stagnalis is a non-native species of lakes and brackish tidal rivers. It is infrequent in CT where it is potentially **invasive** and prohibited. Native to Europe, the plant can spread by seed or rooted stem fragments and form dense colonies. It does not grow in highly acidic waters.

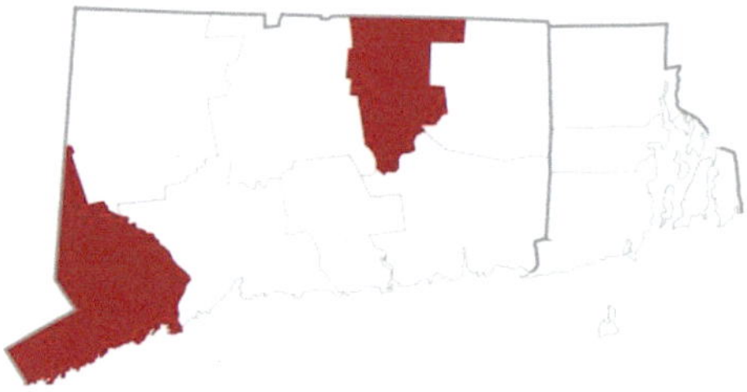

Ceratophyllum — Ceratophyllaceae

Hornwort

The Hornworts, or Coontails, are submerged, perennial, highly branched plants that float or are suspended in the water column. Plants are rootless but may become anchored to a substrate if segments become buried. Leaves are in whorls (of 3–11 leaves) along the stem, dichotomously divided, and forked into linear segments. Inconspicuous, submersed, unisexual flowers are borne in leaf axils. Fruits, if produced, are flattened, smooth-surfaced, dark green or reddish, achenes with spiny margins.

Two species occur in CT and RI, both native. *Ceratophyllum demersum* is one of the most common aquatic plants in the region.

KEY TO SPECIES

1. Leaves usually forked 1–2 times and conspicuously serrated, ultimate segments somewhat flattened; fruits with 3 long spines……………………………………………… *C. demersum*
1. Leaves usually forked 3–4 times, with few marginal teeth, ultimate segments threadlike; fruits with 4+ long and short spines…………………………………………… *C. echinatum*

COMMON HORNWORT *Ceratophyllum demersum* L.

Common hornwort is a highly branched, submersed plant with long (up to 3m) stems and whorled leaves. Leaves (10–30 mm long) are bright green, usually forked 1–2 times into somewhat flattened segments, curving upward, and more crowded at tips of branches. Leaf margins are conspicuously serrated with teeth, each arising from a raised green base. Flowers are small, at the leaf axils. Fruits are flattened and disk-like (3–6 mm long) with 3 long (0.5–12 mm) spines.

HABITAT & DISTRIBUTION

Ceratophyllum demersum is a native species in ditches, marshes, lakes, ponds, and slow streams. It is widespread and common in CT and RI where it can become abundant. It has a relatively stiff habit with coarsely textured leaves, maintaining its shape when out of water. This makes the stems somewhat brittle whereby they can detach, disperse, and form clonal populations.

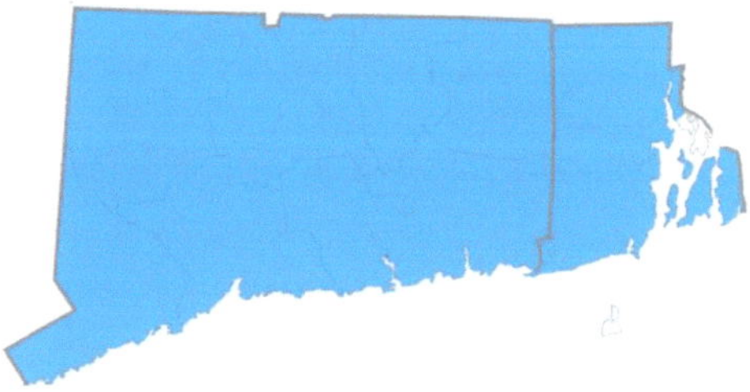

Spineless hornwort *Ceratophyllum echinatum* A. Gray

Spineless hornwort is a highly branched submersed plant with long stems (to 1 m) and leaves in whorls. Leaves are dark green, usually forked 3–4 times into threadlike segments, and not crowded at tips of branches. Leaf margins with few, weak, inconspicuous teeth. Flowers are small, at the leaf axils. Fruits are flattened, disk-like (4.5–6 mm long), and ornamented with 3 long (1–5 mm) spines and 2–13 shorter spines between the longer spines

Habitat & Distribution

Ceratophyllum echinatum is a native species in lakes, ponds, bogs, and slow streams. It is widespread but infrequent in CT and RI. It prefers acidic, low nutrient, shaded, waters. This plant is somewhat stiff, yet not as stiff as *C. demersum*.

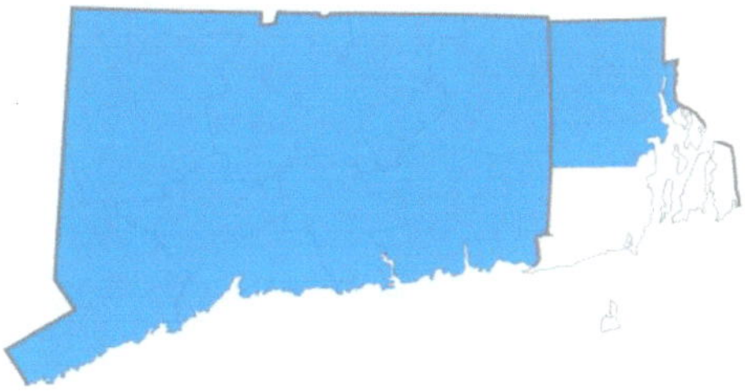

Crassula — Crassulaceae
Pygmy-weed

WATER PYGMY-WEED *Crassula aquatica* (L.) Schönl.

Water pygmy-weed is a diminutive, rooted, annual species. Leaves are opposite, fleshy, linear to lance-shaped (2–6 mm long), pointed at tip, with leaf bases fused to form a boat-shaped cup at each node. Flowers are minute (to 2 mm), with 4 narrow petals, borne singly in a leaf axil. Fruit an oblong follicle.

HABITAT & DISTRIBUTION

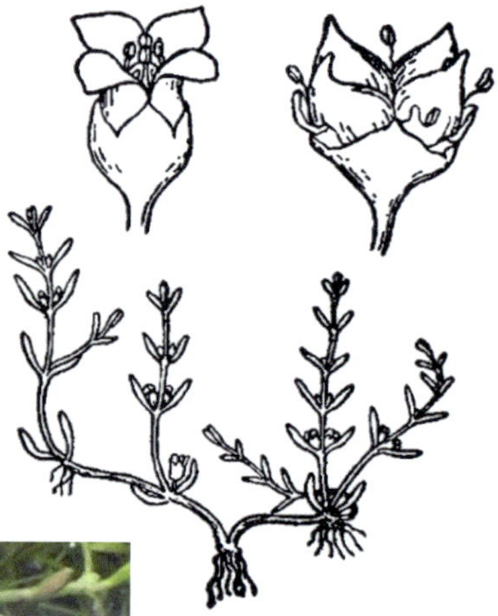

Crassula aquatica is a native, tufted or mat-forming, species of open, coastal intertidal mudflats of brackish to freshwater streams, non-tidal rivers, or less commonly in ponds. It is rare and imperiled in CT, where it is state-listed as ***endangered***.

Notes: In addition to seeds, this annual can disperse by rooted stem fragments. Interestingly, it is one of only four obligate aquatic species of an otherwise xerophytic genus of 200+ species. It is similar to plants of Waterwort (*Elatine*), but is distinguished by its fleshy leaves, fused leaf bases, and follicle fruits.

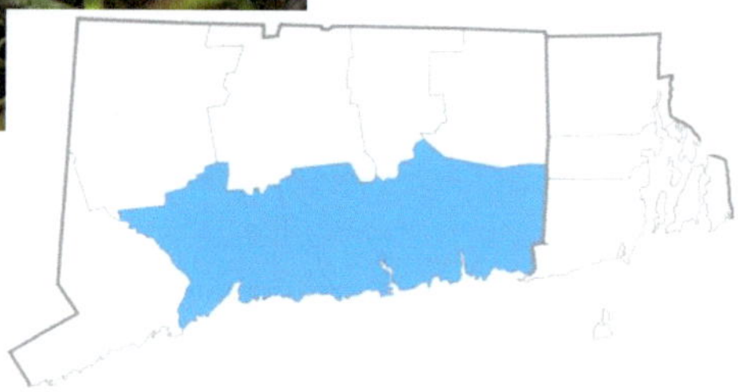

Egeria — Hydrocharitaceae
South American waterweed

BRAZILIAN WATERWEED *Egeria densa* Planch

Brazilian waterweed, or Brazilian elodea, is a submersed, rooted perennial with elongated (to 3 m), flexuous, branched stems. Leaves are in crowded, overlapping whorls of 4–6, lance-shaped (1.5–4 cm long, 3–6 mm wide), bright to dark green, recurved, with finely tooted margins (under magnification). Flowers are unisexual (on separate plants), white, 20 mm in diam., and borne singly from upper leaf axils on slender stalks reaching water surface. Only male plants are known in the US.

HABITAT & DISTRIBUTION

Egeria densa is a non-native species of lakes, ponds, and sluggish rivers. It is uncommon in CT and RI where it is **invasive** and prohibited. The species is native to southeastern Brazil and introduced through the aquarium trade. In the US it reproduces strictly asexually via fragments, grows rapidly, and can outcompete native plants. Can be confused with Waterweeds (*Elodea*).

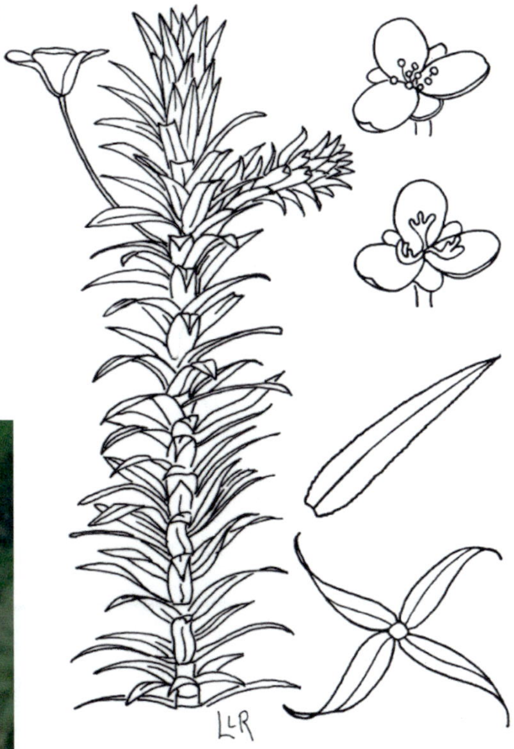

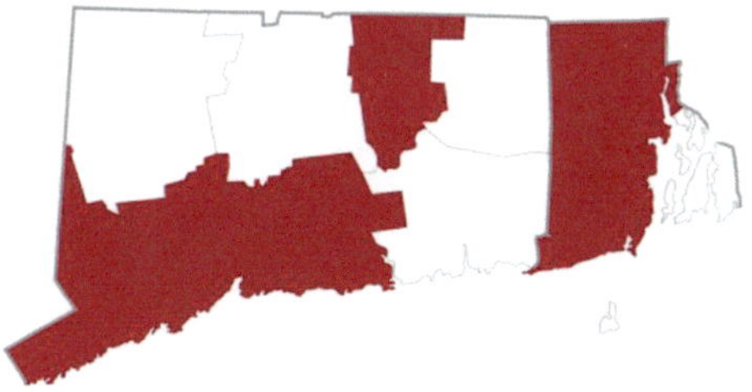

Eichhornia — Pontederiaceae
Water hyacinth

COMMON WATER HYACINTH *Eichhornia crassipes* (Mart.) Solms.

Common Water hyacinth is a free-floating perennial with a buoyant cluster of spongy leaves above water surface and long, feathery, black roots hanging in water column. Leaves are in a basal rosette, green, glossy, and usually with greatly inflated stalks (3–33 cm long). Leaf blades are oval to round (2.5–11 cm long, 4–12 cm wide). Flowers are large (5–7 cm wide) and showy, blue or purplish-blue with a yellow spot, clustered in 4–15-flowered spikes, borne on erect stems (5–12.5 cm long) above the leaves.

HABITAT & DISTRIBUTION

Eichhornia crassipes is a non-native species. It is infrequent in CT and RI, found where plants escaped cultivation or were introduced during the same season. The species cannot tolerate winter in our area.

Notes: Native to South America, this is one of the most aggressive aquatic weeds known in warmer climates where it forms dense floating rafts. Its flowers open for one day only—within two hours after sunrise and then wilt by night.

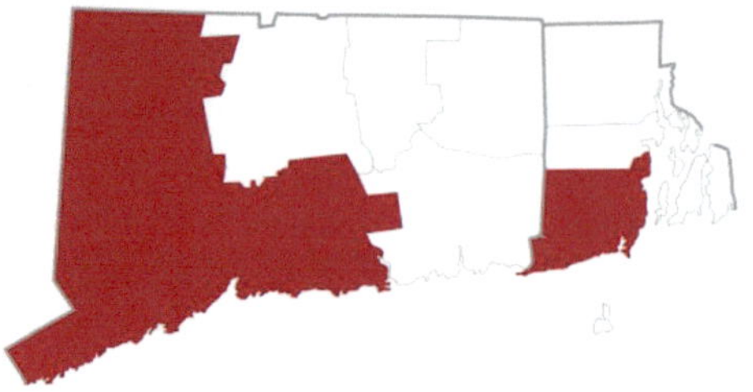

Elatine — Elatinaceae

Waterwort

The Waterworts are diminutive, mat-forming submersed plants, which can easily be overlooked. Individuals are sometimes emersed in mud when stranded by receding waters. Plants have simple, paired leaves along the stems and root at the lower nodes. Single, white or pinkish, bisexual flowers are borne in upper leaf axils which mature into capsules.

Four species are known in CT and RI, including two non-natives. All species are annuals and lack asexual reproduction, so population recruitment is strictly from seeds. Waterworts are popular aquarium plants.

KEY TO SPECIES

1. Leaves oblong and rounded, 1–5 mm long; flowers with 2 petals and stamens; seed surface with rounded pits; plants holding their shape when removed from water.......... *E. minima*
1. Leaves much longer than wide, 3–8 (–15) mm long; flowers with 3 petals and stamens; seed surface with longer 6-sided, angular-ended pits, plants limp and collapsing when out of water..2
    2. Leaf blades obovate or broadly spatulate, the larger blades 1.5–5 mm wide, tips rounded .... *E. americana*
    2. Leaf blades linear to egg-shaped, 0.5–3 mm wide, tips acute or sometimes notched....... .... *E. triandra*

AMERICAN WATERWORT *Elatine americana* (Pursh) Arn.

American waterwort is an erect or prostrate herb with short (1.5–5 cm long), branched stems. Leaves are egg-shaped or broadly spoon-shaped, 3–8 mm long, larger blades 1.5–5 mm wide, with rounded tips. Minute flowers are borne singly in the leaf axils and develop into rounded capsules (1.5 mm wide). Seeds are cylindrical and marked on the surface with 6-sided, angular-ended pits.

HABITAT & DISTRIBUTION

Elatine americana is a native species in shallow (<1 m) waters and muddy shores of ponds, tidal or nontidal rivers. It is rare in RI where it is state-listed as a species of ***concern***.

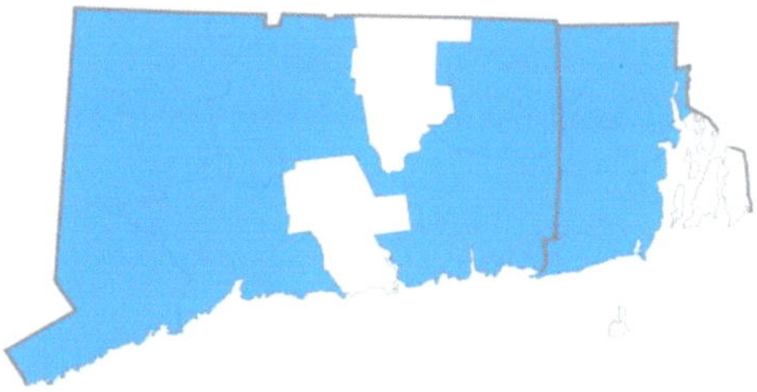

SMALL WATERWORT *Elatine minima* (Nutt.) Fisch. & CA Mey.

Small waterwort is a small, erect, herb with short (to 3 cm long), sometimes few-branched, stems. Leaves are rounded and small (1–5 mm long). Minute flowers are borne singly in the leaf axils and develop into papery capsules containing 4–9 seeds. Seeds are cylindrical and marked on the surface with 8–9 rows of rounded pits.

HABITAT & DISTRIBUTION

Elatine minima grows in shallow (1–2 m) waters of lakes, ponds, tidal or nontidal streams, and exposed mudflats. It is native and widespread in CT and RI. Flowers are self-pollinating, and seeds are dispersed by water. Colonies can become very dense (up to 2500 plants per m^2) reaching highest densities in shallow water.

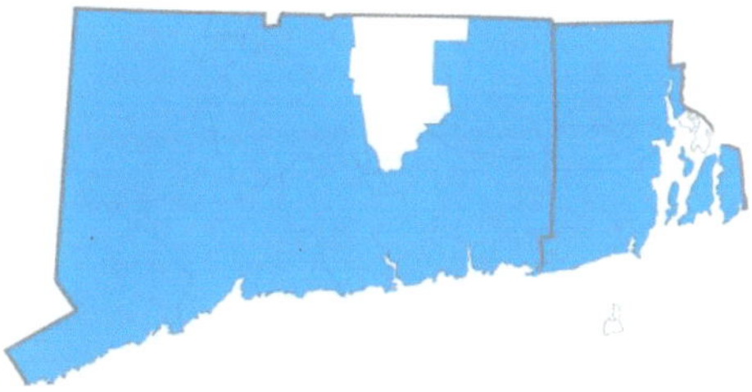

LONG-STEM WATERWORT *Elatine triandra* Schkuhr

Long-stem waterwort is a prostrate herb with short (2–10 cm), highly branched, bright yellow-green stems. Leaves are linear to egg-shaped, 3–8 (–15) mm long, 0.5–3 mm wide, with acute or sometimes notched tips. Minute flowers are borne singly in the leaf axils and develop into papery capsules. Seeds are oblong and marked on the surface with 6-sided, angular-ended pits.

HABITAT & DISTRIBUTION

Elatine triandra is a non-native species in shallow waters (to 1 m) of lakes, ponds, tidal or nontidal streams, and mudflats. It is uncommon, or at least poorly documented, in CT. It is reported from the Connecticut River in MA. Introduced from Eurasia, where it is a weed in rice fields, this species is now widespread throughout North America. It can form dense carpets on lake bottoms and its seedbanks can be persistent. **Notes**: It is very similar to Asian Waterwort (*E. ambigua* Wight), a non-native species, which has become introduced—most likely through aquarium disposal—into New England. Distinguishing between these two species is nearly impossible when plants are found submersed. When emersed, *E. ambigua* produces longer (0.5–2.5 mm), recurved fruit stalks (fruit stalks in *E. triandra* are ≤ 0.5 mm and erect). Recent molecular data confirm *E. ambigua* is present in waters of Middlesex Co., CT, representing one of the first known eastern North American occurrences

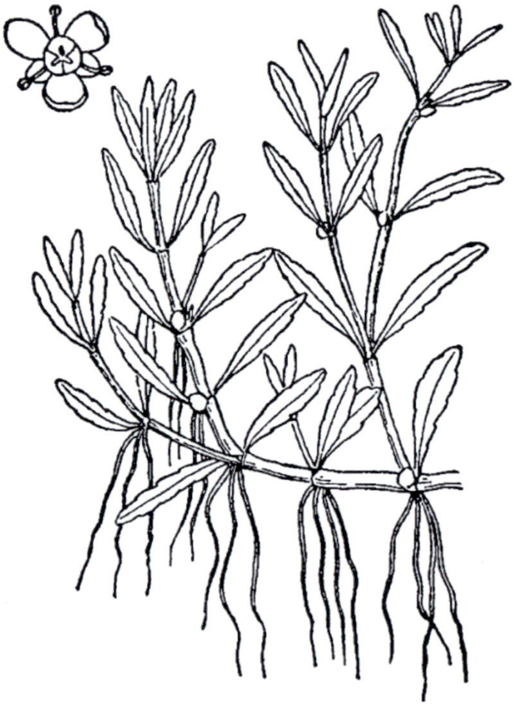

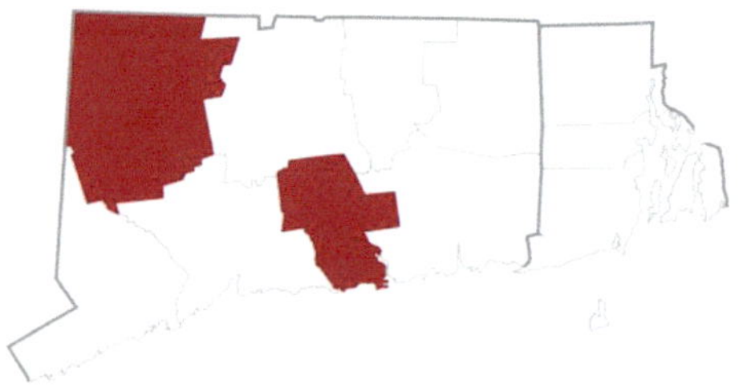

Eleocharis — Cyperaceae

Spikesedge

The Spikesedges, or spikerushes, are rooted, grass-like plants, with green upright stems and very reduced, inconspicuous leaves (represented by a closed, thin sheath at base of stems). Flowers are tiny, borne above water, in a single cluster (spikelet) of overlapping scales at the tip of a stem. Fruits are small and nut-like with bristles at the base, and a distinctive tubercle at top.

Many species of Spikesedge occupy shallow marshy wetlands and wet mud in CT and RI. These species are more wholly emergent (and not detailed here). Three species, all native, can be found more submerged at times in CT and RI.

KEY TO SPECIES

1. Stem much narrower than the spikelet, stems generally round in cross-section, flexible, slender (<0.5 mm), typically short (5–8 cm tall), and often forming a dense turf................2
1. Stem nearly as thick as the spikelet, flowering stems triangular in cross-section, stiff, 16–70 cm tall.. *E. robbinsii*
   2. Plants of freshwater ponds and lakes (rarely brackish), without root-tubers at ends of rhizomes; nutlet with longitudinal rows of fine horizontal ridges .... *E. acicularis*
   2. Plants of brackish or saline coastal tidal marshes, with curved root-tubers at ends of rhizomes; nutlet without longitudinal rows of fine horizontal ridges .... *E. parvula*

NEEDLE SPIKESEDGE *Eleocharis acicularis* (L.) Roem. & Schult.

Needle spikesedge is a delicate, rooted, perennial with white, threadlike rhizomes. Upright stems are clustered, flexible, short (5–8 cm tall), slender (<0.5 mm wide), and narrower than the spikelet. Spikelet is small (2–8 mm long) with dark brown scales with green mid-rib. Nutlet is 3-angled, with 2–4 bristles, and a tubercle separated from body by narrow constriction.

HABITAT & DISTRIBUTION

Eleocharis acicularis is a native species of shallow lake or river shores. It is common and abundant in CT and RI. Plants can grow in dense clusters and form a turf. In deep water, plants may not produce flowers/fruits and can grow much taller (30+ cm).
Note: It closely resembles Little-headed spikesedge of coastal saline or brackish tidal waters which are stouter and have curved, root-tubers on ends of the rhizome.

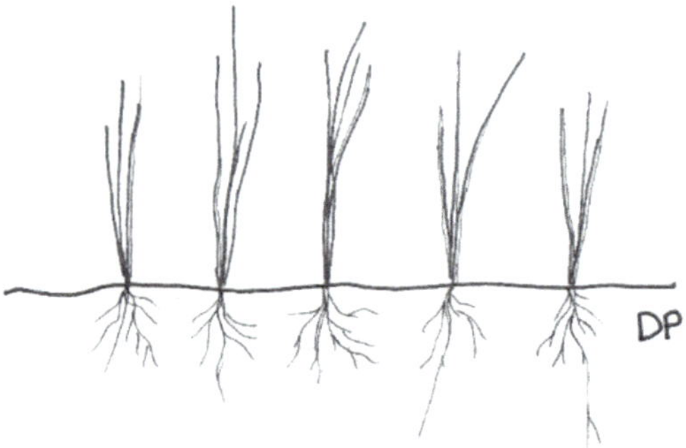

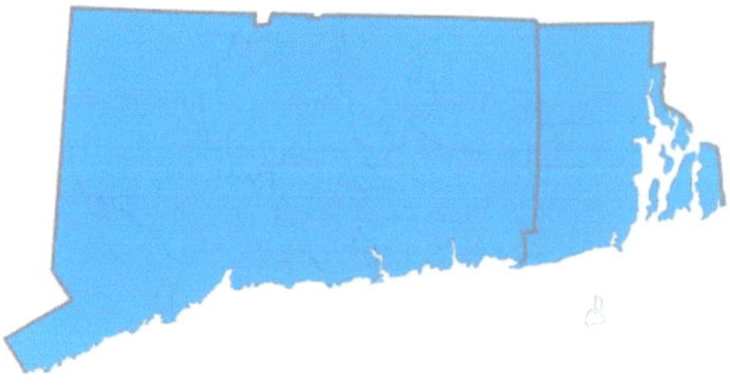

LITTLE-HEADED SPIKESEDGE *Eleocharis parvula* (Roem. & Schult.) Link

Little-headed spikesedge is a delicate, rooted, perennial with slender rhizomes terminated by pale, oblong, usually markedly curved, tubers (2–5 mm long). Upright stems are clustered, flexible, short (5–8 cm tall), and narrower than the spikelet. Spikelet is small (2–4 mm long) with light brown scales. Nutlet is 3-angled, lacking longitudinal rows of fine horizontal ridges, with 5–7 bristles, and a tubercle continuous with the fruit body.

HABITAT & DISTRIBUTION

Eleocharis parvula is a native species of coastal brackish or saline tidal waters in CT and RI. In deep water, plants may not produce flowers/fruits.
Note: It closely resembles Needle spikesedge of more freshwater habitats, which lacks root-tubers at ends of rhizomes.

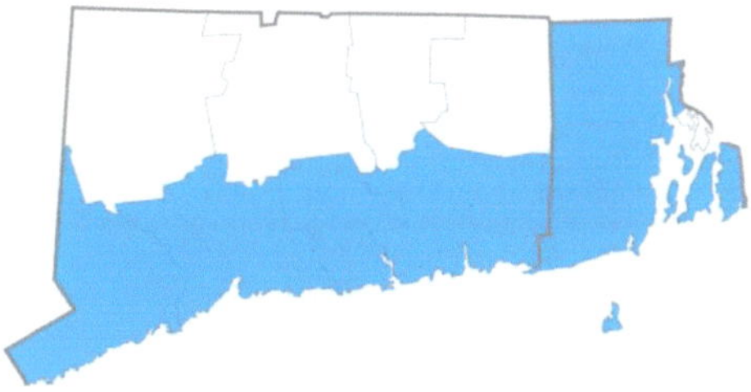

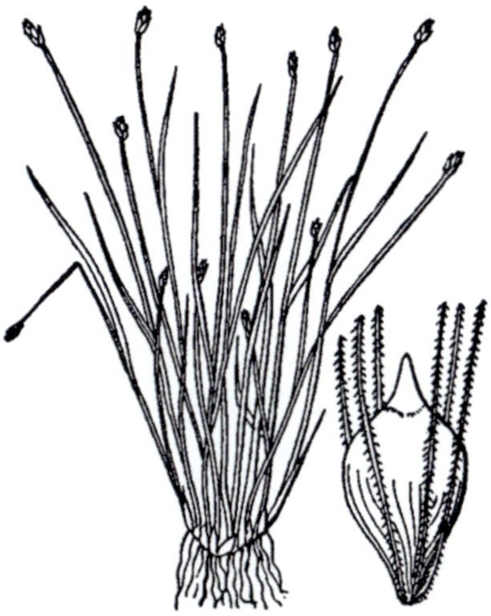

Robbins' spikesedge is a rooted, perennial with buried, reddish, rhizomes (1–2 mm thick). Flowering stems are stiff, erect (16–70 cm tall, 1–2 mm wide), 3-sided, and nearly as thick as the spikelet. Numerous submersed, thread-like, flexible stems (resembling leaves) lacking spikelets are frequently present attached to base of plant. Spikelet is 9–33 mm long**.** Nutlet is 2-sided, with 6–7 bristles, and a distinct tubercle.

HABITAT & DISTRIBUTION

Eleocharis robbinsii is a native species of lakes and ponds where it forms clumps. It is widespread in CT and RI.

Note: It is similar to Water bulrush (*Schoenoplectus subterminalis*). Water bulrush is distinguished by alternate leaves and flowering stems with only their top portions held above water which have a single bract (2–20 mm long) attached below flower cluster appearing like a continuation of the stem.

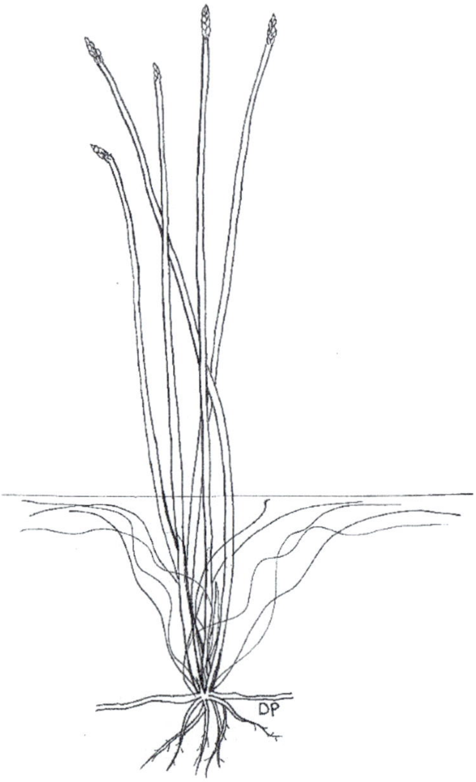

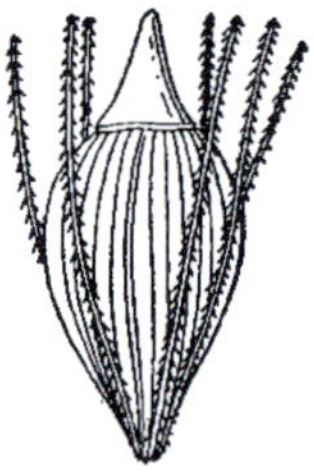

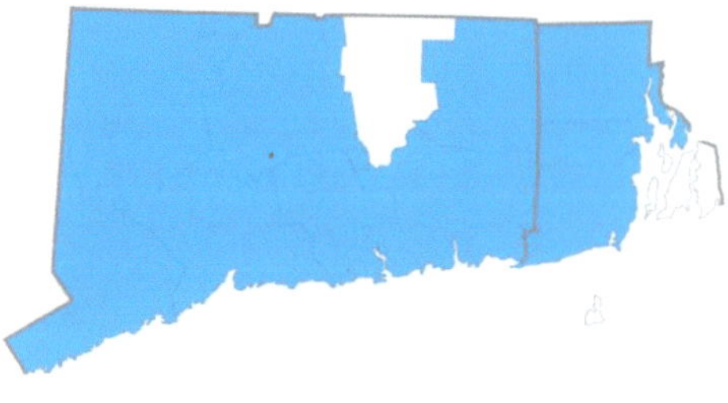

Elodea — Hydrocharitaceae

Waterweed

The Waterweeds are rooted, submersed aquatics with elongated, slender, branching stems which are somewhat brittle. Plants are leafy with small (<15 mm long, <5 mm wide), stalkless, bright green, recurved leaves borne in whorls of 2–3 per node. Flowers are unisexual (on different plants), white, and borne at the water surface.

Two species, both native and similar vegetatively, are known from CT and RI. Commonly sold as aquarium "oxygenating" plants.

KEY TO SPECIES

1. Leaves oblong, mostly 2 mm wide (ranging 1–5 mm wide), firm, blunt-tipped, densely crowded and overlapping towards stem tips.. *E. canadensis*
1. Leaves narrowly lance-shaped, mostly 1.3 mm wide (ranging 0.9–1.7 mm wide), weak, pointed, more or less equally spaced along stem.. *E. nuttallii*

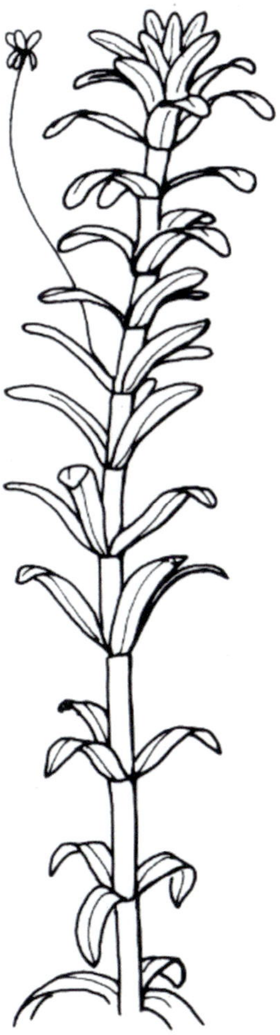

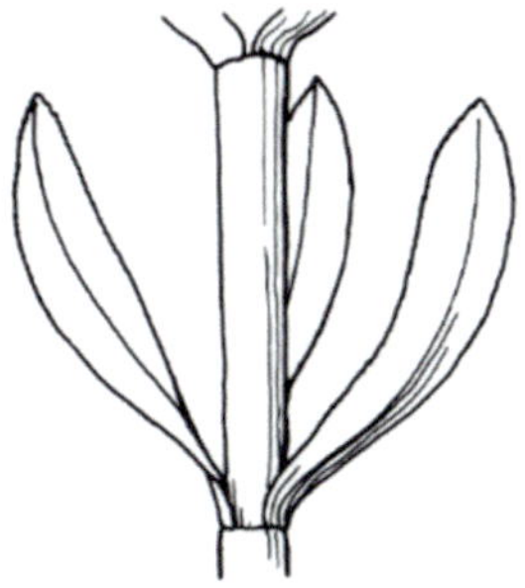

COMMON WATERWEED *Elodea canadensis* Michx.

Common waterweed is a submersed, perennial with elongated (≥1 m), flexuous stems rooting from nodes. Leaves are mostly in whorls of 3, oblong, blunt tipped, mostly 2 mm wide (6–10 mm long, 1–5 mm wide), firm, and flat. Leaves in upper whorls can grow closely together and overlap. Flowers are unisexual (on separate plants), small (petals 2–5 mm long), borne singly, and white. Female flowers attached to long thread-like stalks at leaf axils, with sepals 2–3.5 mm long. Male plants are uncommon, but have flowers also elevated on long (2–30 cm) stalks.

HABITAT & DISTRIBUTION

Elodea canadensis is a native species of lakes, ponds, and slow rivers. It is widespread in CT and RI but most common in neutral to calcareous waters. Plants grow rapidly in favorable conditions and form dense masses. It spreads by the detachment of resilient, densely leaved, stem fragments (turions).

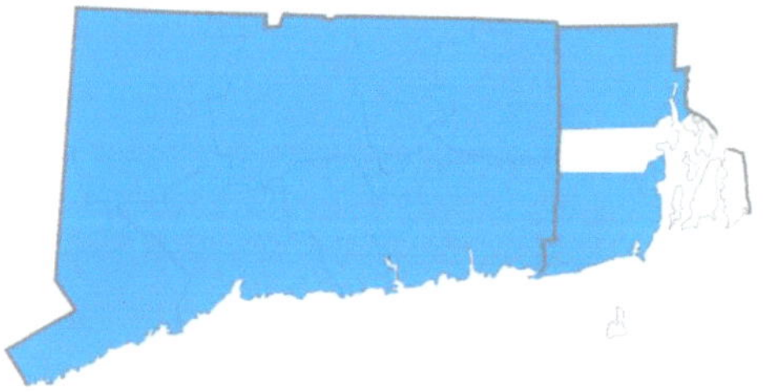

FREE-FLOWERED WATERWEED *Elodea nuttallii* (Planch.) St. John

Free-flowered waterweed is a submersed, perennial, with elongated, flexuous stems rooting from nodes. Leaves are mostly in whorls of 3, narrowly lance-shaped, pointed, mostly 1.3 mm wide (4–15 mm long, 0.9–1.7 mm wide), weak, and edges slightly folded along midvein. Leaves more or less equally spaced along stem. Flowers are unisexual (on separate plants), small (petals 2.5 mm long), borne singly at leaf axils, and white. Female flowers attached to long thread-like stalks, with sepals 1–1.8 mm long. Male flowers are stalkless and released from plant to float on water surface.

HABITAT & DISTRIBUTION

Elodea nuttallii is a native species of lakes, ponds, slow rivers, and sometimes brackish tidal waters. It is widespread and common in CT and RI. Plants are found in acidic to moderately alkaline waters.

Eriocaulon — Eriocaulaceae

Pipewort

The Pipeworts, or Hatpins, are diminutive, submerged, tufted, erect perennials with conspicuously white roots bearing distinctive cross-walls. Leaves are green, in dense, basal rosettes, narrow, tapered to a point from a flattened, pale, spongy base. Leaves are smooth, veiny, and somewhat translucent. Soft, flowering heads of minute, whitish flowers are borne at tips of erect, slender, longitudinally ridged stalks, and held above water surface.

Two species occur in CT and RI. Both are native and very similar in appearance.

KEY TO SPECIES

1. Flower cluster 4–6+ mm wide, white to gray-white, globose in shape, atop a 5–7 ridged stalk; usually a single flowering stalk per plant... *E. aquaticum*
1. Flower cluster 3–4 mm wide, dull gray or rarely yellow-brown, hemispherical in shape, atop a 4–5 ridged stalk; 1–4 flowering stalks per plant... *E. parkeri*

SEVEN-ANGLED PIPEWORT *Eriocaulon aquaticum* (Hill) Druce

Seven-angled pipewort is a tufted, submersed plant with basal, green, tapered leaves, 1–10 cm long (to 40 cm), 2–5 mm wide. Flowering stalks (1 mm wide), usually one per leaf rosette, with 5–7 vertical ridges. Flower cluster 4–10 mm wide, white to gray-white, and globose in shape.

HABITAT & DISTRIBUTION

Eriocaulon aquaticum is a native plant of freshwater lakes, ponds, and slow rivers. It is widespread and common in CT and RI. Plants are usually abundant on sandy substrates of shallow, low nutrient, acid waters.

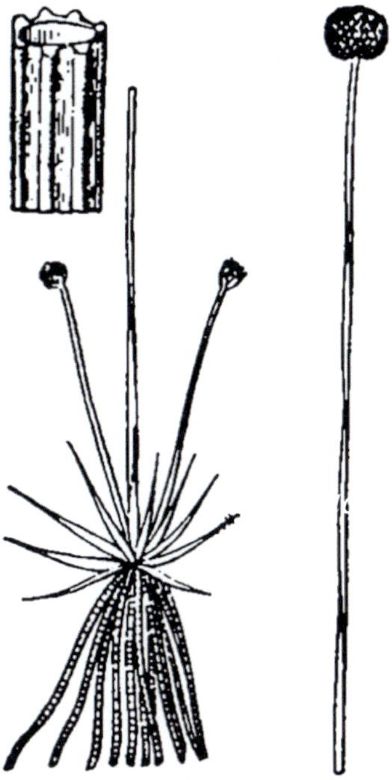

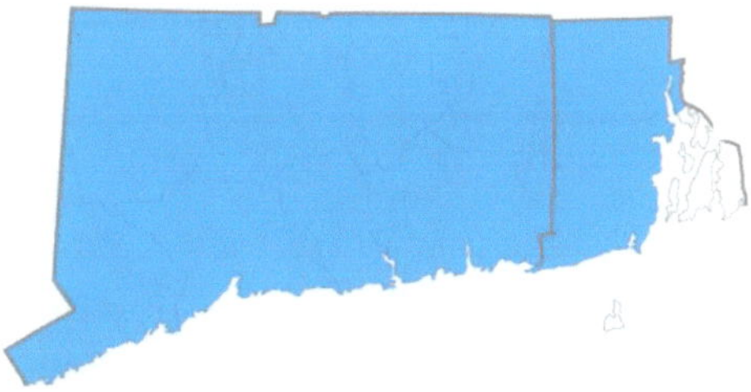

ESTUARY PIPEWORT *Eriocaulon parkeri* B.L. Robins.

Estuary pipewort, or Parker's pipewort, is a tufted submersed plant with basal, green, tapered leaves, 2–6 cm long, 1–2 mm wide. Flowering stalks, 1–4 per leaf rosette, are 2.5–10 cm long, with 4–5 vertical ridges. Flower cluster 3–4 mm wide, dull gray or rarely yellow-brown, and hemispherical in shape.

HABITAT & DISTRIBUTION

Eriocaulon parkeri is a native species of fresh to brackish tidal shores and open mudflats of estuaries. Restricted to coastal regions and rare in CT where it is state-listed as ***endangered***.

Notes: Plants are usually exposed at low tide, found at intertidal reaches to shores above high tide, and tolerant of oligohaline (salinity 0.5–5.0 ppt) waters.

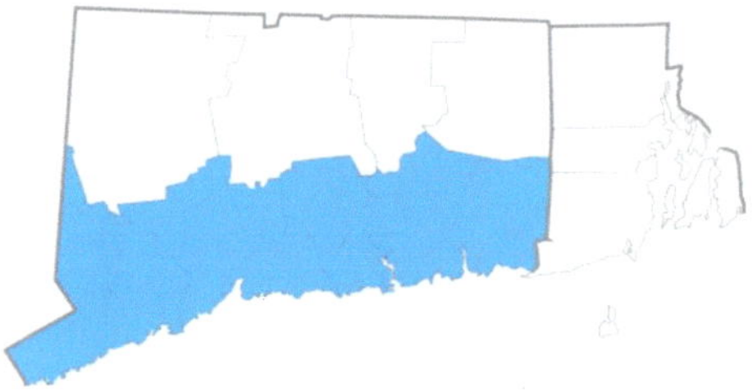

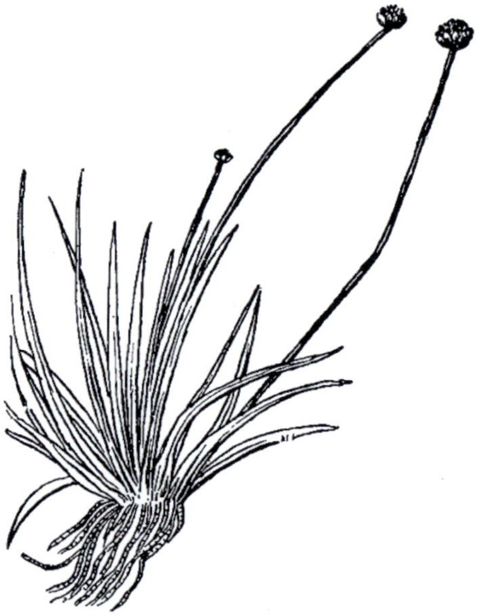

Glossostigma — Phrymaceae
Mud-mat

MUD-MAT *Glossostigma cleistanthum* W.R. Barker

Mud-mat is a diminutive (1–2 cm tall), mat-forming, submersed, perennial herb with creeping, slender rhizomes, rooting at the nodes (nodes to 18 mm apart). Leaves are opposite in pairs, erect, small (9–25 mm long, 1–3 mm wide), strap-shaped but widening toward tip. Flowers are small, with 2 stamens, borne singly on short (0–2 mm) stalks between the leaf pairs. When submersed, flowers are closed and self-pollinating. Fruit a capsule (<2 mm).

HABITAT & DISTRIBUTION

Glossostigma cleistanthum grows in shallow to deep (4 m) waters, or on exposed shorelines, of lakes, ponds, and streams. This Australia native was recently introduced and established in the northeast—presumably via improper aquarium disposal—where it is known in both CT and RI. It thrives in acidic, nutrient-poor waters with full sun. This fast-spreading ***invasive*** weed can form dense, carpet-like colonies (>25,000 plants/m^2), potentially threatening native hydrophytes. It can be confused with Atlantic Mudwort (*Limosella australis*).

Notes: Two ecotypes (submersed or emersed) exist: if growing emersed on open shores, plants exhibit shorter leaves (5–11 mm long), shorter internodes (1–10 mm), longer flower stalks (>4 mm), and behave as annuals with open, insect-pollinated flowers.

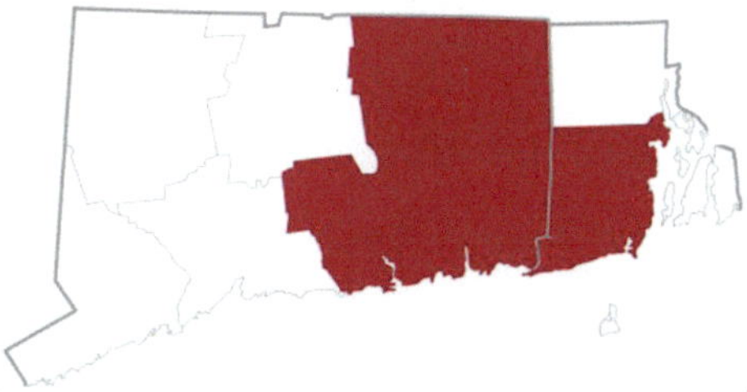

Glyceria — Poaceae
Manna grass

NORTHERN MANNA GRASS *Glyceria borealis* (Nash) Batch.

Northern manna grass, or Small Floating manna grass, is a rooted, perennial grass (60–100 cm tall) with creeping rhizomes and narrow (1.5–5 mm), ascending, hairless stems. Leaves are typically floating on surface, especially early in season (becoming weakly erect with age). Leaf blades are linear (15–45 cm long, 2–7 mm wide), flat, abruptly tapered to a point, hairless, dark bluish-green, with a prominent midvein. Leaf blades are tubular at base and sheath the stem (sheaths, deep red when submerged, have fused margins for most of their length). Flower clusters are emergent, weakly erect (18–40 cm tall), several-branched, and borne from the uppermost leaf sheath. Flowers are enclosed in small (10–20 mm), linear, spikelets.

HABITAT & DISTRIBUTION

Glyceria borealis is a native grass of shallow lake and stream margins. It is widespread in CT. Mid-stem leaf blades have a dense covering of minute, pimply projections on their upper surface making them strongly hydrophobic.
Note: It can be confused with other widespread species, like Eastern manna grass (*G. septentrionalis* A.S. Hitchc.) and Sharp-scaled manna grass (*G. acutiflora* Torr.). These two plants (both excluded here) differ mostly by finer details of flowers.

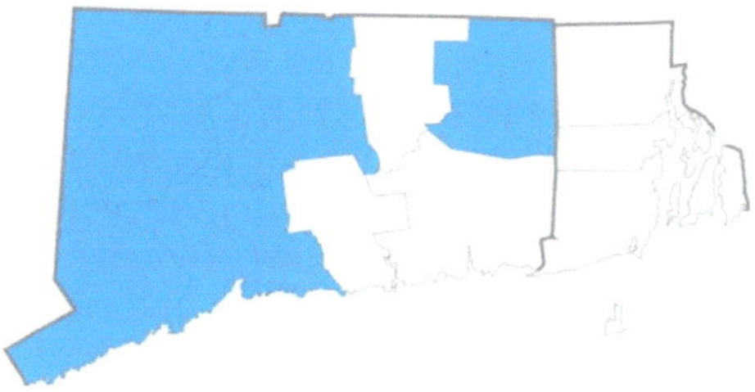

Gratiola — Plantaginaceae
Hedge-hyssop

GOLDEN HEDGE-HYSSOP *Gratiola aurea* Pursh.

Golden hedge-hyssop, or Golden pert, is a rooted, submersed, perennial with slender rhizomes. Plants are diminutive (1–5 cm tall), somewhat fleshy/stout, with smooth, somewhat four-angled, upright stems. Leaves are opposite, stalkless, small (≤5 mm long), narrow, awl-like and pointed, with toothless margins. Plants of this submersed life-form are sterile (vegetative only).

HABITAT & DISTRIBUTION

Gratiola aurea is a native species of low nutrient, mostly acidic, lake waters with full sun exposure. It is widespread in CT and RI, but most common on the Coastal Plain. Plants can be locally abundant, forming dense, turf-like, beds in either shallow or deep (4 m) waters.

Notes: The dwarfed, truly aquatic life-form (f. *pusilla* Fassett) is described here. At many occurrences, the species instead exists as a larger, fertile, emergent life-form on wet sandy shores of marshes, lakes and rivers. The emergent form has taller stems (10–30 cm) with gland-tipped hairs, larger leaves (1–3 cm long) with conspicuous glandular dots, and bright yellow flowers (12–16 mm long) borne singly in leaf axils. The submersed form is similar in appearance to Waterwort (*Elatine*) which has rounded leaves and evident flowers/fruits.

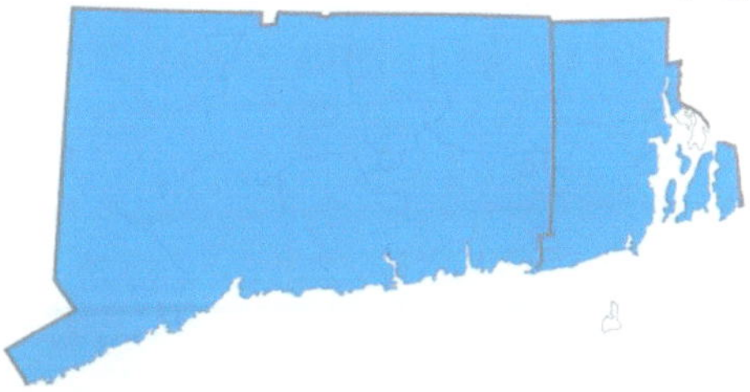

Helanthium — AlisMataceae
Burhead

DWARF BURHEAD *Helanthium tenellum* (Mart.) Britt.

Dwarf burhead is a diminutive (5 cm tall), inconspicuous, submersed, rooted annual. Leaves are basal, submersed, and lance-linear in shape (1–6 cm long, 2–3 mm wide). If emersed on shore, leaves have long (1–9.5 cm), ridged stalks and obvious blades. Flowers are bisexual, small (6–8 mm wide), white, stalked (5–30 mm), and borne in erect clusters.

HABITAT & DISTRIBUTION

Helanthium tenellum is a native of wet, sandy pond shores, with marked seasonal water level fluctuations. This globally rare species is known in CT where it is state-listed as ***endangered*** (considered *State* ***historical*** from RI). It is cultivated as an ornamental aquarium plant (under Pigmy chain sword or *Echinodorus tenellus* (Mart.) Buch.).

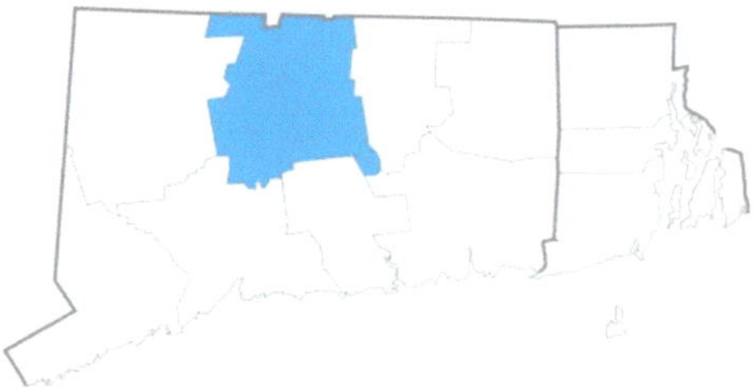

Heteranthera — Pontederiaceae

Mud-plantain

The Mud-plantains are submerged perennials or annuals with elongated stems rooted in mud. Leaves are linear and submerged or with broad blades and floating, with sheathing bases. Flowers are conspicuous, yellow or white, with 6 petal-like segments, 3 stamens, and borne above water. Fruits are elongated capsules.

Two species occur in CT, both native. Individual flowers are short-lived and only bloom for one day.

KEY TO SPECIES

1. Leaves ribbon-like and mostly submersed; flowers pale yellow and borne singly...... *H. dubia*
1. Leaves with stalked, kidney-shaped blades and floating (or emersed); flowers white, borne in clusters of 2-8.. *H. reniformis*

H. dubia

GRASS-LEAVED MUD-PLANTAIN *Heteranthera dubia* (Jacq.) MacM.

Grass-leaved mud-plantain, or Water star-grass, is a submersed (or emersed on exposed mud), rooted, perennial with ribbon-like leaves, buried horizontal stems, and elongated, flexible, branching stems growing to water surface. Leaves are alternate, submersed, linear (3–10 cm long, 1–5 mm wide), sheathing at base, thin, lacking a distinct mid-rib, and blunt-tipped. Flowers are yellow, with six linear (4–11 mm long) petal-like lobes, three yellow stamens, and borne singly at water surface.

HABITAT & DISTRIBUTION

Heteranthera dubia is found in ponds, lakes, and the Connecticut River. It is native to CT and most common in hard waters, uncommon elsewhere. Plants exhibit most extensive growth in alkaline waters and can become weedy.

Notes: Flowers open within 2 hours after dawn and wilt by dusk. When not flowering, this species resembles narrow-leaved Pondweed (*Potamogeton*) species but lacks the conspicuous midvein of the latter.

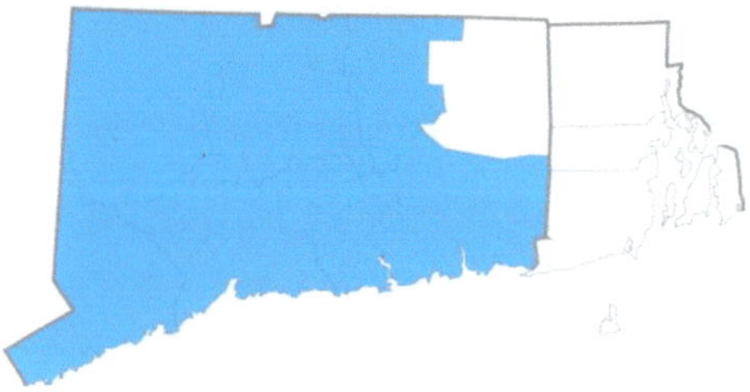

KIDNEY-LEAVED MUD-PLANTAIN *Heteranthera reniformis* Ruiz & *Pavón*

Kidney-leaved mud-plantain is a submersed, annual or short-lived perennial, with elongated, flexible, branching stems growing upright in water. Leaves are alternate, mostly floating (or emersed) with broad, kidney-shaped blades (1–4 cm long, 1–5 cm wide), long (2.5–4 cm) stalks, and stipules (1–5 cm long), and submersed, thin, linear (2.5–4 cm long, 3–8 mm wide) leaves in basal rosettes. Flowers are white, with six petal-like lobes (3–6.5 mm long), borne in clusters of 2–8 above water surface.

HABITAT & DISTRIBUTION

Heteranthera reniformis is found in tidal river mudflats characterized by fluctuating water depths. It is native and it reaches its northern limits in CT where it is state-listed as ***special concern*** (considered historic/extirpated).

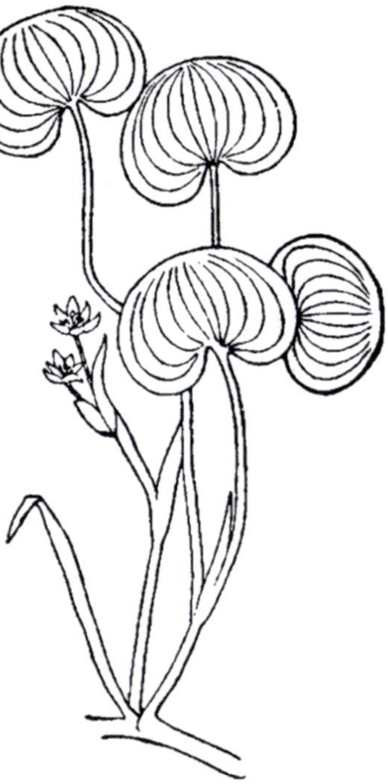

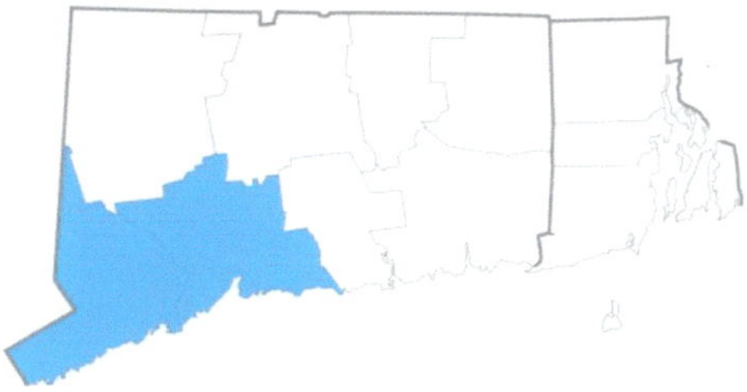

Hottonia — Primulaceae
Featherfoil

AMERICAN FEATHERFOIL *Hottonia inflata* Ell.

American featherfoil is a submersed, short-lived annual with long (30–50 cm), flaccid, hollow stems, rooted at base. Leaves are alternate (or whorled/clustered near top), submersed (or floating at surface), oblong (2–8 cm long), feathery, and finely divided into linear segments. Flower clusters are floating on water surface, with swollen, air-filled, spongy stalks (to 30 cm long) bearing whorls of (3–10) inconspicuous white flowers. Fruit is a globose (1.5–3 mm) capsule.

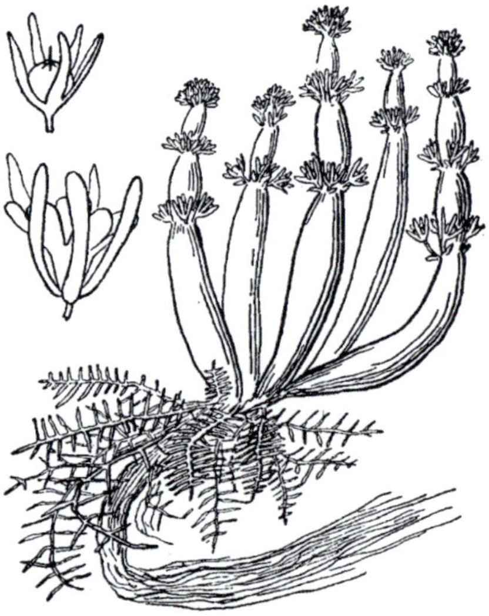

HABITAT & DISTRIBUTION

Hottonia inflata is a native Coastal Plain species found in still, acidic waters of ponds, swamps, and ditches. It is rare and state-listed as a ***concern*** in CT and RI.

Note: This odd-looking plant is unmistakable in spring with its buoyant, swollen, flower clusters. These clusters can detach and become free-floating. As a "winter annual" species, plants typically wither and die by mid-July after fruiting. Abundance at a site can vary greatly from year to year as a function of precipitation.

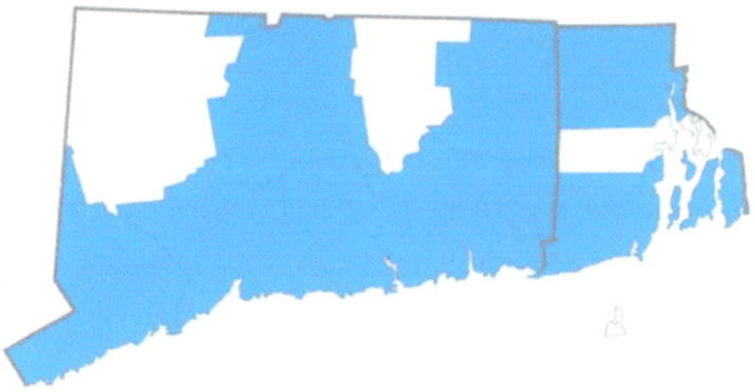

Hydrilla – Hydrocharitaceae
Hydrilla

HYDRILLA — *Hydrilla verticillata* (L.f.) Royle

Hydrilla, or Water-thyme, is a submersed, perennial with rhizomes and elongated, flexuous, branched stems. Leaves are in whorls of 4–8, linear (8–15 mm long, 1–4 mm wide), sharply pointed, with saw-toothed margins, and bottom surface with minute prickles or bumps along midvein. Flowers are unisexual (on same or separate plants), small, white, and borne singly from upper leaf axils. Female flowers with elongated (10–50 mm) floral tubes reaching water surface. Fruit a small (4–6 mm long) capsule.

HABITAT & DISTRIBUTION

Hydrilla verticillata is a non-native species of lakes and rivers. It is known in CT, especially problematic around Connecticut River, where it is **invasive** and prohibited. This Asian species was introduced into the US as an aquarium plant. It produces two types of overwintering turions: underground, pale brown, tuber-like ones and dark green, stiff-scaled ones on upright stems.

Notes: A problematic weed that is fast-growing, easily and rapidly dispersed, and with remarkably broad ecological tolerances (of water depth, pH, salinity, and nutrient levels). Anywhere it is found it poses a serious threat to the waterbody. It may be confused with Waterweeds (*Elodea*).

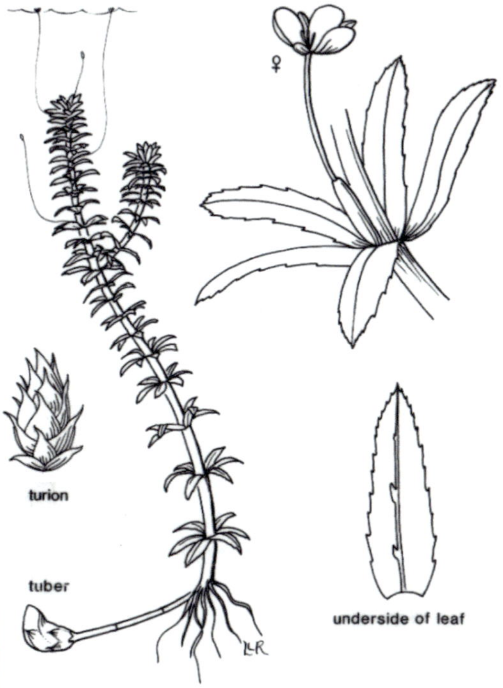

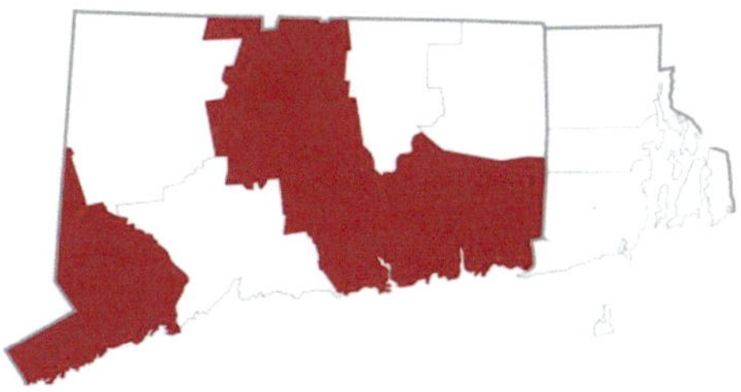

Hydrocotyle — Apiaceae
Water pennywort

MANY-FLOWERED MARSH PENNYWORT *Hydrocotyle umbellata* L.

Many-flowered marsh pennywort is a small, perennial, floating-leaved plant with slender, creeping stems rooting at the nodes. Leaves are floating (or emersed if stranded), one per node, with thick, green, nearly-round blades (10–75 mm) and round-toothed margins. Leaves have long stalks (to 25 cm) centrally attached to underside of blade. Flowers clusters (3 cm wide), one per node, are borne at tip of long stalks. Flowers (15–35 per cluster), connected at same point on stalk, are minute and whitish. Fruits are round (2–3 mm wide), ribbed, and dry.

HABITAT & DISTRIBUTION

Hydrocotyle umbellata occurs in shallow (<1 m) pond shores, bogs, and ditches. It is rare and state-listed as ***endangered*** in CT. This native species reproduces vegetatively by stem elongation and detached stem fragments, and has broad ecological tolerances.

Notes: Similar to the rare, state-listed Whorled marsh pennywort (*H. verticillata* Thunb.) of coastal CT (*Historic* in RI). The rare species produces flowers in spike-like clusters, each with multiple (2–12) whorls of flowers, and wider fruits (3–4 mm).

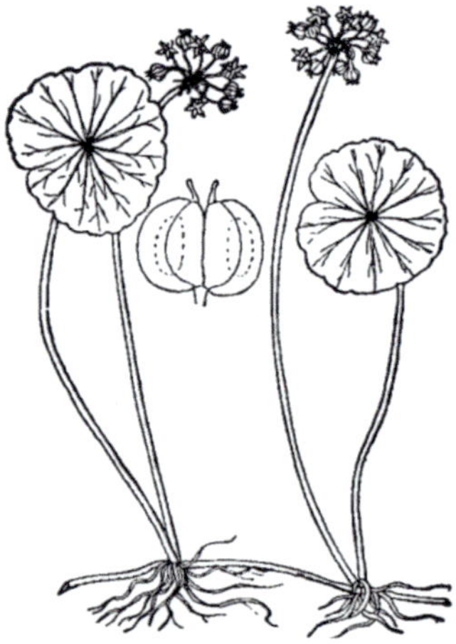

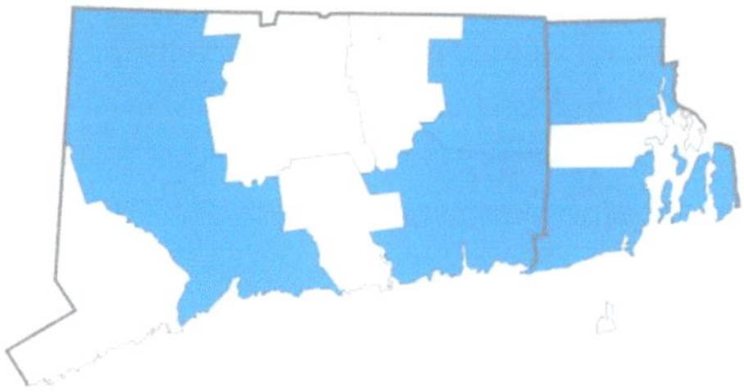

Hypericum — Hypericaceae
St. John's-wort

NORTHERN ST. JOHN'S-WORT — *Hypericum boreale* (Britt.) E. Bickn.

Northern St. John's-wort is a delicate, submersed, rooted, perennial, with slender rhizomes and flaccid, elongated, mostly unbranched stems (1–2 mm wide). Leaves are opposite, small (6 mm long), rounded, toothless, 3-veined, and without stalks (round bases of blade clasping stems). Plants usually lack flowers.

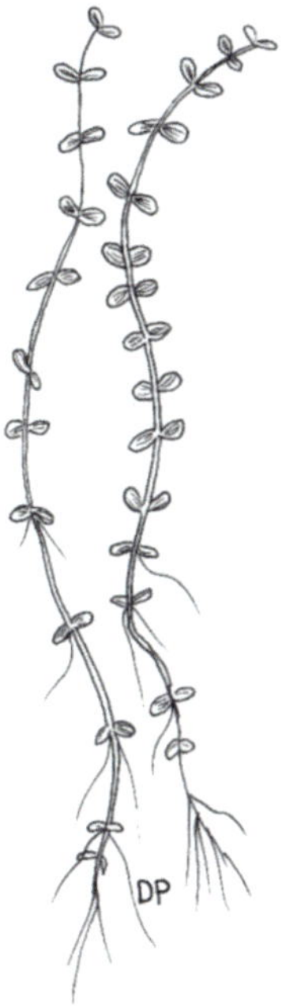

HABITAT & DISTRIBUTION

Hypericum boreale is a native species that occurs in shallow ponds, streams, and marshes. It is widespread in CT and RI.
Notes: The flexuous, sterile, submersed life-form (f. *callitrichoides* Fassett) is described here. In many cases, the species also occurs nearby as the typical, fertile, emergent life-form on wet shores or marshes. Emergent plants have erect stems (10–40 cm tall), larger, more oblong, leaf blades (1–3 cm long) with scattered translucent dots on surface and 3–5 veins, and small, yellow flowers borne near tips of spreading upper branches. The submersed life-form is similar to Water starworts (*Callitriche*).

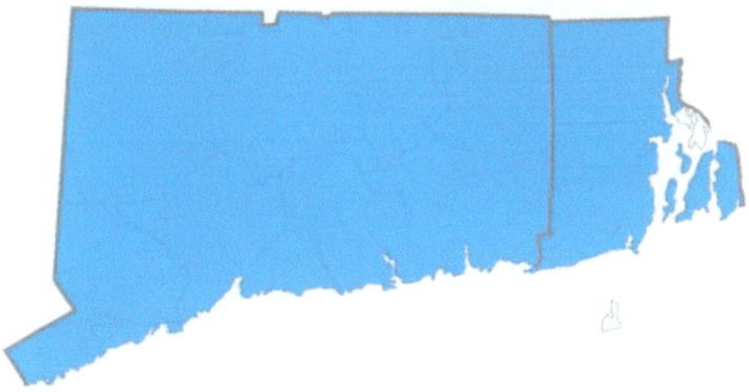

Isoetes — Isoetaceae

Quillwort

The Quillworts are tufted, submersed, perennial, spore-producing (i.e., non-flowering) plants with grass-like leaves arranged in rosettes, and rooted below. Leaves are linear, hollow, green, with scattered cross-walls, and swollen, whitish bases where the spore-producing structures (sporangia) are enclosed.

Five species occur in CT and RI, all native. Vegetatively, most species are very similar in appearance. Species identification is challenging and requires microscopic examination of sculptural details of mature macrospores. These diagnostic, larger, spherical spores (bone-white when dry), usually borne in bases of outer leaves, are typically available from our plants only after July.

KEY TO SPECIES

1. Megaspores densely covered with thin, spine-like projections; leaves deciduous........................ *I. echinospora*
1. Megaspores covered with ridges; leaves lasting 1+ years........................2
   2. Megaspore diameter averaging <0.5 mm, patterned with an unbroken network of ridges........................ *I. engelmannii*
   2. Megaspore diameter averaging >0.5 mm, patterned with crest-like, wrinkled or network of broken ridges........................3
3. Leaves bright green with abundant stomates; plants occasionally emergent; megaspore girdle (area just below equatorial ridge) obscure........................ *I. riparia*
3. Leaves dark green to red-brown with few or no stomates; plants always submerged; megaspore girdle evident........................4
   4. Megaspore diameter averaging >0.6 mm; leaves coarse and firm, abruptly tapering to the tip; plants typically in deep (1–3+ m) water........................*I. lacustris*
   4. Megaspore diameter averaging <0.6 mm; leaves soft and flaccid, gradually tapering to the tip; plants typically in shallower (0.5–1 m) water........................ *I. tuckermanii*

I. echinospora

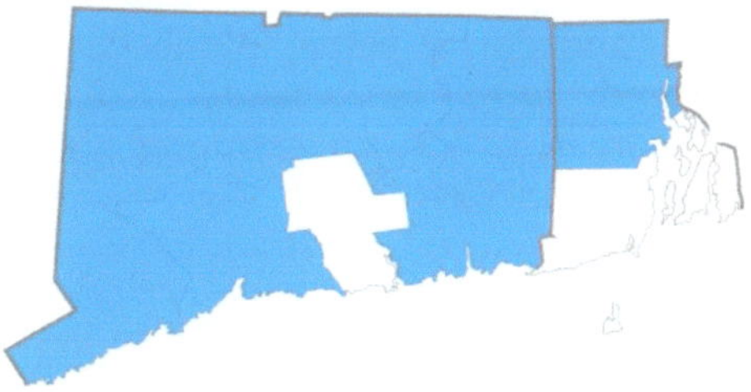

Leaves are soft, mostly 5–15 cm (to 40 cm) long, 0.5–1.5 mm wide, with slender tips. The more common and widespread species in CT, but rare and state-listed as ***concern*** in RI, occupying clear, low nutrient, acidic lakes, ponds, streams. **Note:** Our plants are spp. *muricata* (Durieu) A. & D. Löve. It is known to hybridize with *I. engelmannii* in CT.

ENGELMANN'S QUILLWORT *Isoetes engelmannii* A. Braun

Leaves are 6–50 cm long, 0.5–2 mm wide. Usually in shallow waters of ponds and slow streams. Plants can be emersed later in season. It is widespread in CT but rare and state-listed as ***concern*** in RI.

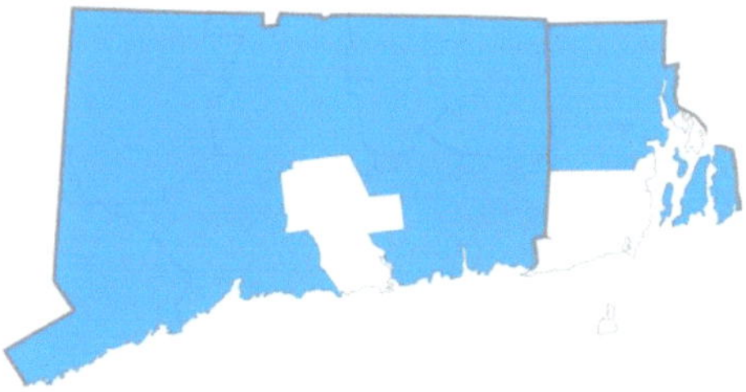

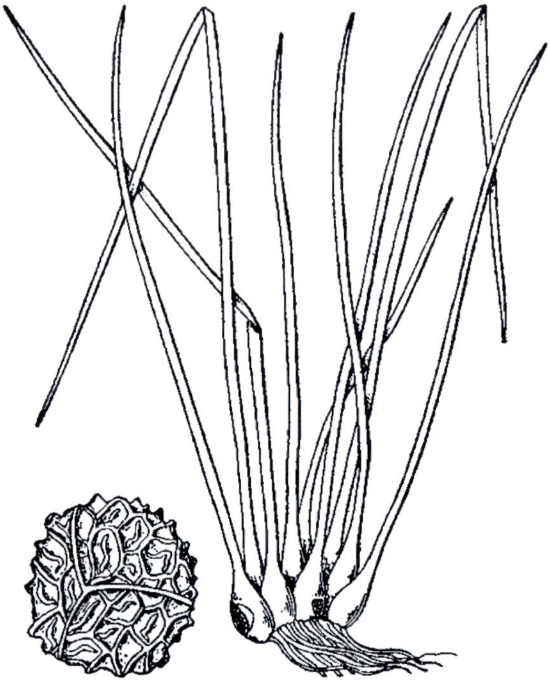

LAKE QUILLWORT *Isoetes lacustris* L.

Leaves are dark green, coarse and firm, short (5–10 cm long, 1–2 mm wide), abruptly tapering to the tip, with few or no stomates. Plants are submerged in deep (1–3+ m), cold, clear lakes, ponds and streams. It is reported from Fairfield Co., CT.

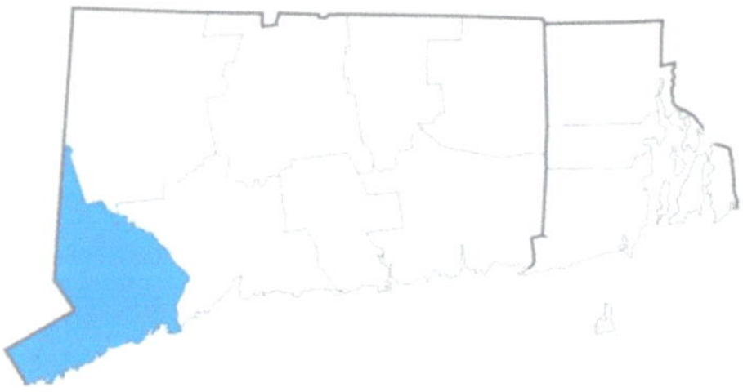

RIVERBANK QUILLWORT *Isoetes riparia* Engelm. ex A. Braun

Leaves are bright green to yellow-green, 10–20 cm long, 0.5–1.5 mm wide, with abundant stomates. Plants usually in shallow waters (<0.5m deep), occasionally emergent, of streams and lakes, including brackish tidal rivers. It is uncommon in CT and rare and state-listed as ***concern*** in RI.

Note: Our plants are var. *canadensis* Engelm.

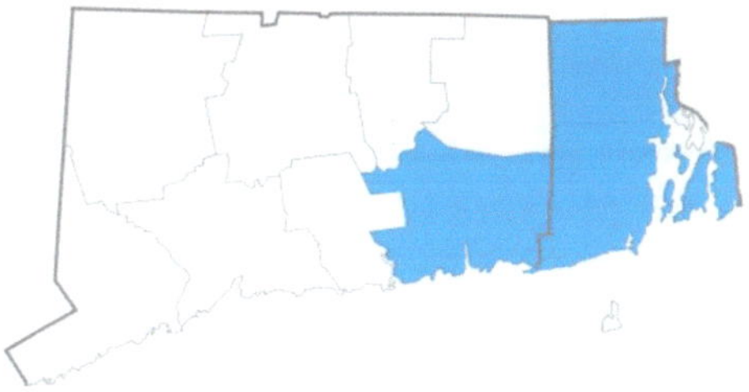

TUCKERMAN'S QUILLWORT *Isoetes tuckermanii* A. Braun ex Engelm.

Leaves are thin, soft and flaccid, mostly 5–20+ cm long, 0.25–1.5 mm wide, gradually tapering to the tip. Plants always submerged, in ponds and lakes 1 (–3) m deep. It is widespread in CT and RI.

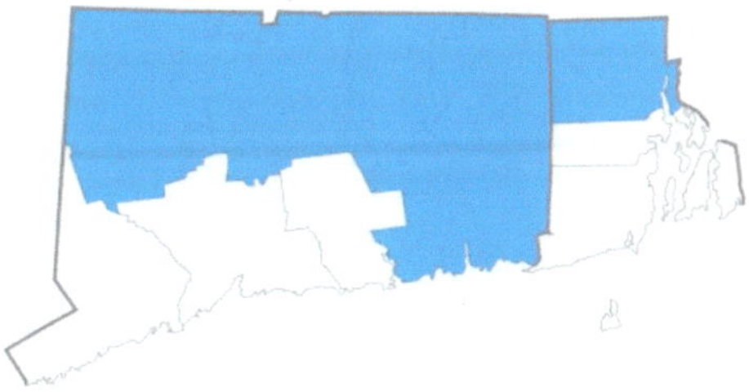

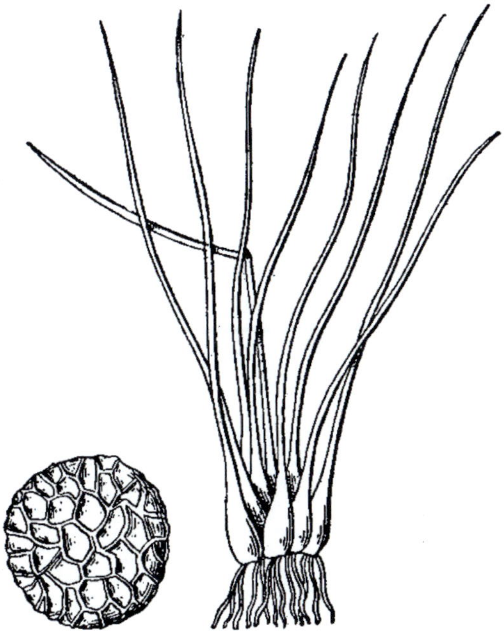

Juncus – Juncaceae
Rush

BAYONET RUSH *Juncus militaris* Bigelow

Bayonet rush is a rooted, grass-like perennial, with submersed rhizomes and stout, erect stems. Stems are tall (to 150 cm) and round or oval in cross-section (5–12 mm wide). Leaves are of two types: many, submerged (or tips floating), flaccid, thread-like leaves attached to rhizomes, and usually a single, emersed, stiff leaf attached to the erect stem. The emersed leaf is stout (50–100 cm long, 2–4 mm wide), round or oval in cross-section, with hardened internal cross-walls, pointed, attached near middle of stem, and overtopping the flower cluster. Flowers are minute, green or brown, borne in a multi-branched cluster (4–15 cm tall) at stem tip. Fruit is a small (2–3 mm long), 3-seeded capsule.

HABITAT & DISTRIBUTION

Juncus militaris is a native species of shallow lakes and rivers. It is widespread in CT and RI. The submersed leaves are found particularly in deeper waters, where plants may remain sterile. Plants can form dense colonies.

Notes: The hardened partitions in leaves, present at regular intervals, can be felt by pulling the leaf between your fingers and pressing gently. Similar to Robbins' spikesedge (*Eleocharis robbinsii*) and Water bulrush (*Schoenoplectus subterminalis*), but neither possess emersed leaves or internal cross-walls.

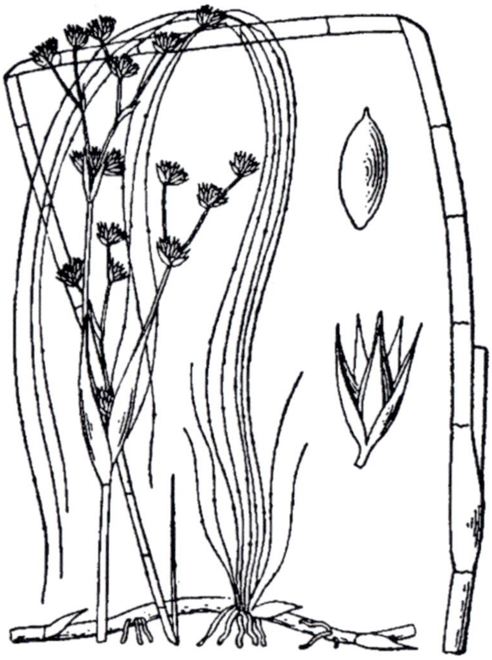

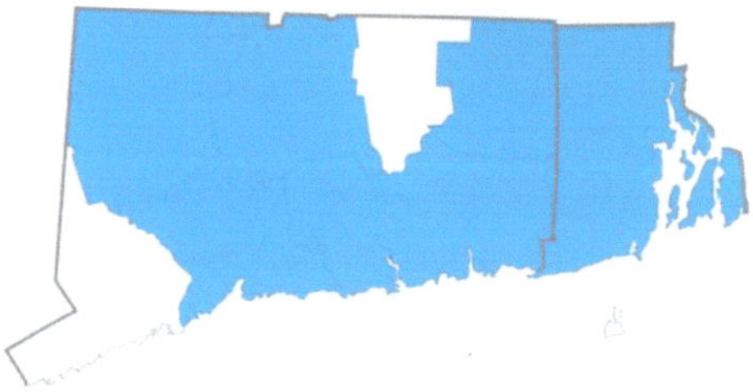

Lemna — Araceae

Duckweed

The Duckweeds are small (<15 mm long), green, perennials, mostly floating on surface of quiet waters. Plant body (or "thallus") is greatly reduced (not differentiated into stem and leaf), flattened, oblong to narrowly egg-shaped, very obscurely 1–5 veined above, and one root below extending into water. Usually 1–several plants are connected. Flowers and fruits are tiny, rarely seen.

Four species occur in CT and RI. Identification is best when plants are fresh. Individuals consistently reproduce vegetatively and are usually abundant, often intermixed with other duckweeds (*Spirodela* and *Wolffia*).

KEY TO SPECIES

1. Plant bodies 6–15 mm long, oblong to broadly lance-shaped in outline, tapering to a conspicuous narrow stalk by which mature plant bodies remain attached laterally to parents, floating just below water surface in tangled masses; root often absent.... *L. trisulca*
1. Plant bodies 1–6 mm long, round, egg-shaped, or narrowly oblong in outline, without stalks, usually floating on surface of water; root present.. 2
  2. Mature plants <0.5–1.5 mm wide, sides somewhat parallel, obscurely 1-veined (may appear unveined).. *L. valdiviana*
  2. Mature plants 1.3–3 mm wide, sides curved, with 3–5 obscure veins..........................3
3. Plants sometimes with small (0.8–1.6 mm wide) green-brown rootless turions; thallus often with reddish coloration on lower surface, usually with a distinct row of minute raised points along the midvein on upper surface, and 3 veins......................................*L. turionifera*
3. Plants without turions; thallus typically green on lower surface, with or without a distinct row of minute raised points along the midvein, and 3–5 veins.............................*L. minor*

Common duckweed — *Lemna minor* L.

Common duckweed is a diminutive, perennial, floating on water surface. Its body, or thallus is small (1–6 mm long, 1–3 mm wide), egg-shaped, with 3–5 very obscure veins on top surface, with or without a row of raised projections along midvein. Root is mostly over 3.5 cm long (to 15 cm), with a rounded tip, extending into water. Turions not produced.

Habitat & Distribution

Lemna minor is a native species of ponds, lakes, marshes and streams. It is common and widespread in CT and RI.

Notes: Similar to Turion duckweed which differs by having a reddish color on lower surface, 3 veins, a distinct row of minute projections along midvein, and sometimes small, rootless turions. Also similar to Minute duckweed (*L. perpusilla* Torr.) which is reported from RI where it is state-listed as ***Concern***. It differs by having much shorter (<3.5 cm long), pointed roots, a body with a prominent, raised projection at tip, and a winged root sheath.

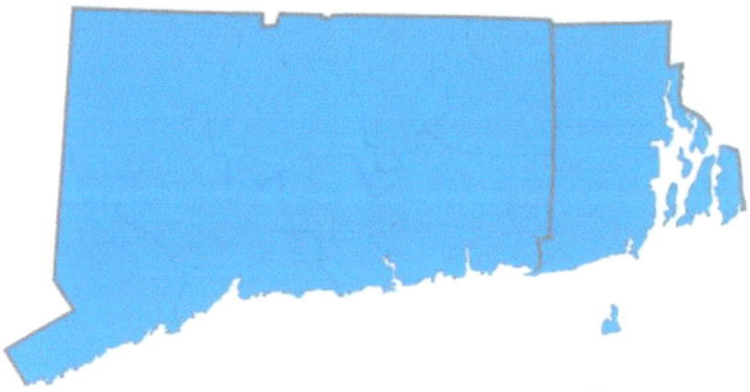

IVY-LEAVED DUCKWEED *Lemna trisulca* L.

Ivy-leaved duckweed is a diminutive perennial, suspended just below water surface. Its body, or thallus, is 6–15 mm long (1.4–4 mm wide), oblong to broadly lanceolate in outline, with a somewhat toothed margin towards tip, tapering at base to a conspicuous, narrow stalk. Root deciduous and sometime absent. Mature plant bodies remain attached to parents via stalks.

HABITAT & DISTRIBUTION

Lemna trisulca is a native species of ponds and lakes, most frequent in alkaline waters. It is somewhat widespread in CT and RI.

Notes: One of the more distinctive duckweeds with its long stalks. This species often forms extensive, tangled colonies where 3–30 clones can be interconnected. Bodies become shorter and thicker when flowering.

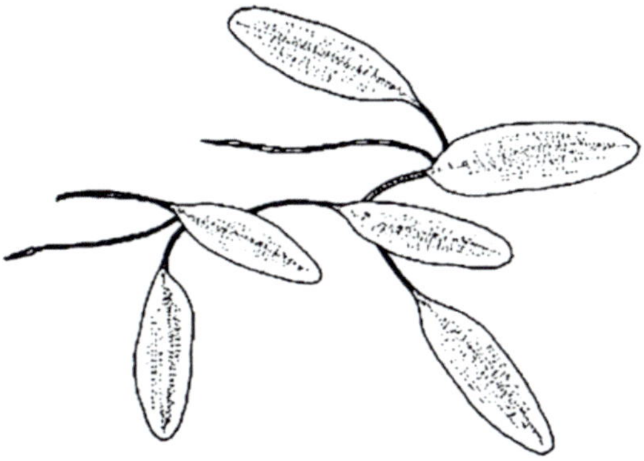

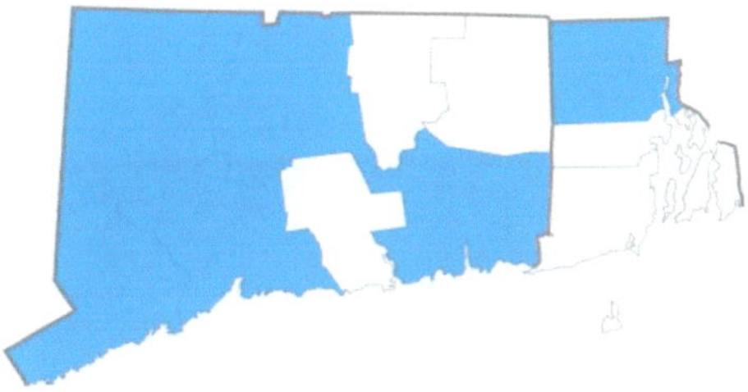

TURION DUCKWEED *Lemna turionifera* Landolt

Turion duckweed is a diminutive, perennial, floating on water surface. Its body, or thallus is small (1–4 mm long, 1–3.5 mm wide), egg-shaped, often with a reddish under surface, with 3 obscure veins on top surface, and usually with a distinct row of raised points along the midvein on upper surface. In unfavorable conditions, plants form small (0.8–1.6 mm wide), olive to brown, rootless turions.

HABITAT & DISTRIBUTION

Lemna turionifera is a native species. It is uncommon (but probably under collected) in CT and state-listed as a species of ***concern*** in RI.

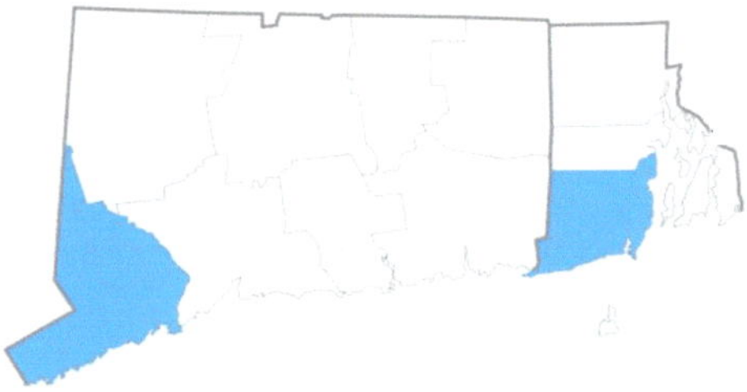

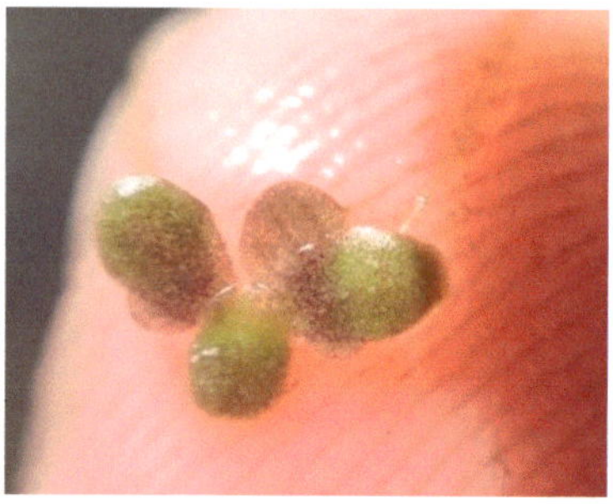

PALE DUCKWEED *Lemna valdiviana* Phil.

Pale duckweed is a diminutive, perennial, floating on water surface. Its body, or thallus is small (2.5–5 mm long, <0.5–1.5 mm wide), oblong, sides somewhat parallel, with 1 vein on top surface, with or without a row of raised projections along midvein. Root to 1.5 cm long, with a rounded to pointed tip, extending into water. Turions not produced.

HABITAT & DISTRIBUTION

Lemna valdiviana is a native species of ponds, lakes, marshes and streams. It is widespread in CT and rare and state-listed as a species of ***concern*** in RI.

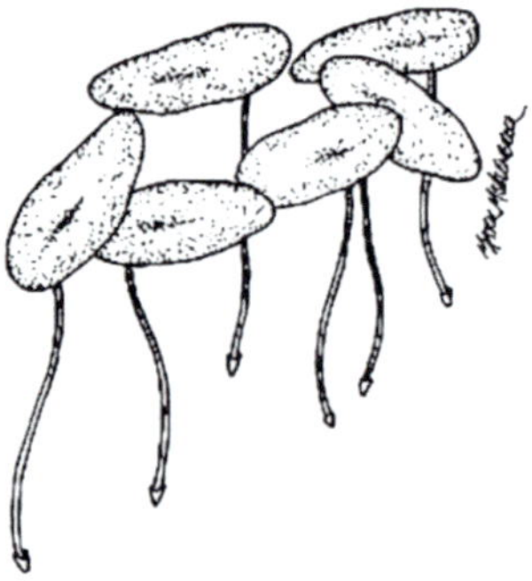

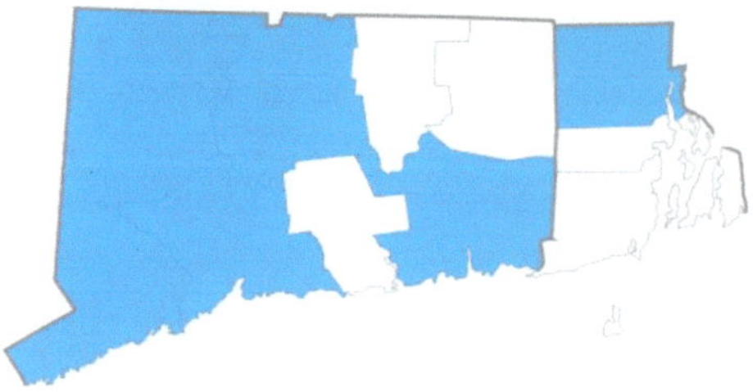

Lilaeopsis – Apiaceae
Grasswort

Eastern grasswort — *Lilaeopsis chinensis* (L.) Kuntze

Eastern grasswort is a low, submersed or emergent, perennial with creeping, horizontal rhizomes which root at nodes. Leaves are 1–few, arising from the rhizome, linear or spatulate in shape, without an expanded blade, 1–5 cm long, hollow, rounded to flattened, transversely divided with 4–6 septa. Flowers inconspicuous, bisexual, and borne 4–10 in clusters from base of leaves. Flower cluster stalks are about as long as or longer than the leaves. Fruits are small (<3mm), rounded nutlets with 5 corky ribs.

Habitat & Distribution

Lilaeopsis chinensis is a native species in shallow waters and exposed mudflats of freshwater to brackish marshes, tidal streams, canals, and ditches or saltmarshes. It is rare in both CT and RI, where it is state-listed as ***special concern*** and ***endangered***, respectively.

Notes: This is a coastal species typically occupying intertidal areas which become inundated at high tide. It can form dense colonies in sunny areas. Submersed plants may remain sterile.

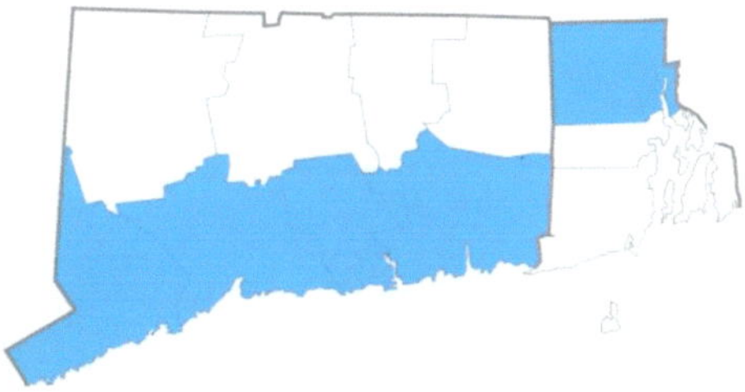

Limnobium — Hydrocharitaceae
Spongeplant

AMERICAN SPONGEPLANT *Limnobium spongia* (Bosc) Steud.

American spongeplant, or American frogbit, is a perennial, free-floating or rooted floating-leaved (or emergent) plant, with horizontal stoloniferous stems and branched roots. Leaves are in basal rosettes and of two forms: floating with short stalks and a thick, broad, spongy pad on lower blade surface and/or emergent with flat, leathery blades on long spongy stalks. Leaf blades are broad (2–8 cm long, 2–5 cm wide), heart shaped to round, (faintly 5–7 veined), usually with pointed tips. Each leaf with a stipule fused to base of leaf stalk. Flowers are unisexual (on same plant), whitish or yellowish, and held above water. Fruits round (4–12 mm wide), berry-like, on recurved stalks.

HABITAT & DISTRIBUTION

Limnobium spongia is a native Coastal Plain species of shallow pond margins. It was known only from central CT, where it may not have persisted. Its occurrence there represented the species' most northeastern locale.

Notes: The conspicuous broad, spongy pad on the undersides of leaf blades is most prominent on young floating leaves (becoming less prominent with age).

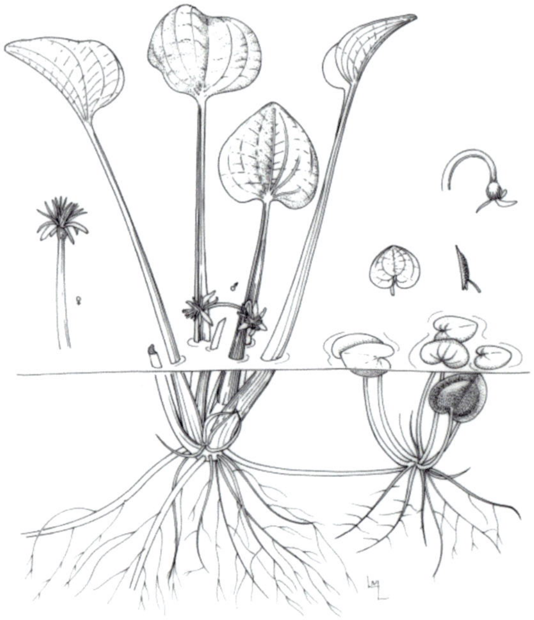

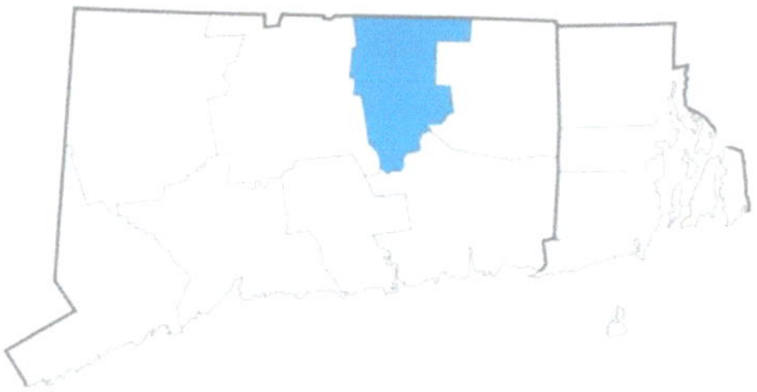

Limosella — Scrophulariaceae
Mudwort

ATLANTIC MUDWORT *Limosella australis* R. Br.

Atlantic mudwort is a diminutive, rooted, submersed, herb with horizontal stoloniferous stems rooting at nodes. Leaves are fleshy, linear (2–5 cm long, 1–2 mm wide), round in cross-section, and tufted, 5–10 leaves clustered at bottom of plant. Flowers are small (3 mm wide), with 5 white petals (3.5–4 mm long) and 4 stamens, borne singly on erect stalks (about half of leaf length) that arise from nodes, several per plant. Fruit is a round capsule.

HABITAT & DISTRIBUTION

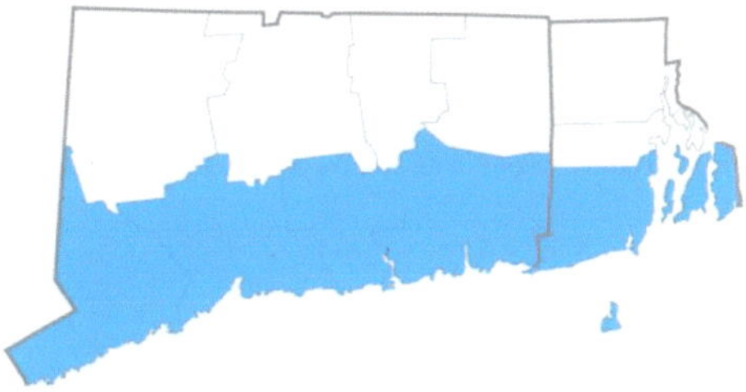

Limosella australis is a native species of shallow, tidally exposed, sand or mud of estuaries, rivers, and ponds. It is rare in both CT and RI, where it is state-listed as ***special concern*** and ***concern***, respectively.

Notes: This succulent, sometimes mat-forming, plant is tolerant of brackish water (to 18 ppt salinity) and exposure at low tides. It is most common and abundant in shallower waters. Technically a perennial species that acts as an annual due to difficulty in surviving winter conditions. It can be confused with Mud-mat (*Glossostigma*).

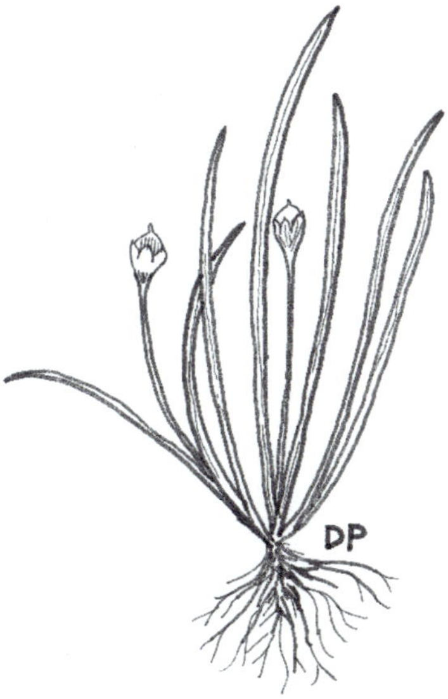

Lobelia — Campanulaceae
Lobelia

WATER LOBELIA *Lobelia dortmanna* L.

Water lobelia is a short, rooted, submersed, perennial plant. Leaves are clustered in basal rosettes, 2–5 cm long, linear, tubular (two hollow tubes in cross section), blunt-tipped, and arching outward. Flowers (1–2 cm long) have white to bluish, united petals, borne loosely towards the end of an emergent stalk. Flower stalks protrude from water surface about 30 cm (but can be 2–3 m long overall). Fruit a green, barrel-shaped, nodding, capsule.

HABITAT & DISTRIBUTION

Lobelia dortmanna is a native species of still, shallow waters of lake shores. It is widespread in CT, but state-listed as a species of ***concern*** in RI. This slow-growing plant prefers soft, clear, nutrient-poor waters with a sandy substrate. It is intolerant of high-nutrient conditions.

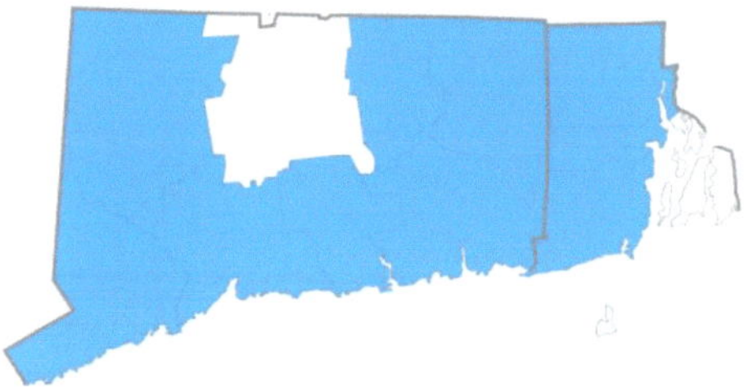

Ludwigia — Onagraceae
Water-primrose

Common water-primrose *Ludwigia palustris* (L.) Ell.

Common water-primrose, or Water purslane, is a mostly submersed (or emergent on mud) perennial, with reddish, many-branched stems (3–50 cm long) laying on sides or partly floating and rooting at lower nodes. Leaves are opposite, green to reddish-bronze, abruptly stalked (5–25 mm long) with egg-shaped to oval, toothless, blades (5–30 mm long, to 20 mm wide). Flowers are inconspicuous, bisexual, borne singly in leaf axils, green-brown, bell-shaped, and topped with 4 triangular sepals. Fruits are oblong (2–5 mm long, 2–3 mm wide), 4-sided capsules with 4 longitudinal green bands.

Habitat & Distribution

Ludwigia palustris is a native species occupying waters of marshes, swamps, rivers and ponds. It is common and widespread in Ct and Ri. Plants can form dense mats in shallow waters.

Note: Similar to the threatened Toothcup [*Rotala ramosior* (L.) Koehne] of gravelly Coastal Plain pond shores which experience seasonal water drawdown. Toothcup, a more emergent species, however, has somewhat 4-sided stems, more lance-shaped leaves, and small white petals.

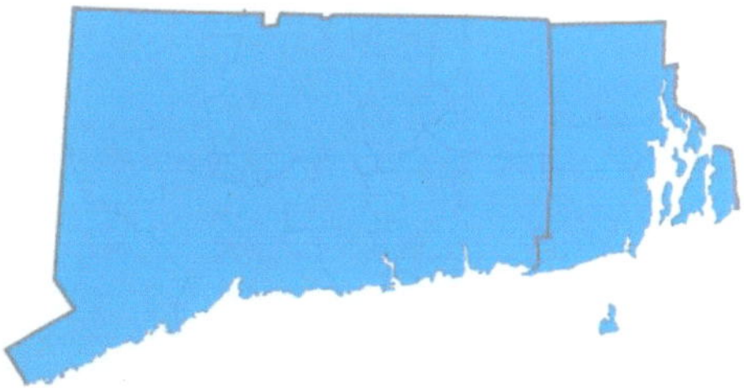

Marsilea — Marsileaceae
Water-clover

EUROPEAN WATER-CLOVER *Marsilea quadrifolia* L.

European water-clover, or Water-shamrock, is a perennial, aquatic fern that grows from long, creeping rhizomes, rooting mostly at nodes. Leaves are usually floating, with blades centrally divided into four triangular segments (each 7–21 mm long, 6–19 mm wide). Leaf stalks are narrow, flexible, and long (6–16 cm). Reproduces by spores, borne in hairy, pea-shaped (5 mm) structures (sporocarps) produced near base of leaf stalks.

HABITAT & DISTRIBUTION

Marsilea quadrifolia is a non-native plant naturalized in shallow pond or slow rivers. It is uncommon in CT where it is considered ***invasive*** and prohibited.

Note: This plant was introduced into New England from Europe**.** The first report of it in North American waters was in 1860 from Bantam Lake (Litchfield Co., CT).

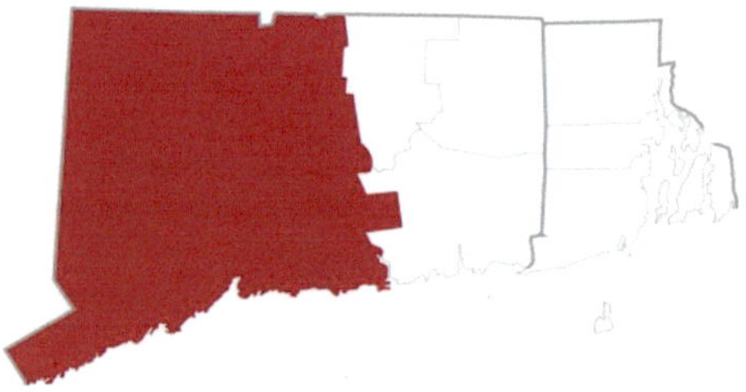

Myriophyllum — Haloragaceae

Water-milfoil

The Water-milfoils are slender, submersed, rooted, perennials with upright stems growing from rhizomes. Leaves are opposite, alternate, or whorled, and mostly feather-like, delicate and finely divided into thread-like segments. The emersed, bract-like, leaves associated with flowers, if present, usually differ in form and are reduced in size. Flowers are unisexual (on same plant in most species), minute, and inconspicuous, borne singly either in leaf axils of submersed leaves or clustered on upright spikes above water. Flowering spikes have male flowers in uppermost positions and female flowers lowermost. Fruits are tiny (to 2.5 mm long), nut-like clusters of four 1-seeded segments.

Ten species occur in CT and RI, including three rare and three non-native. A few species are aggressive and highly invasive. Most species are highly variable so reproductive structures and existence of winterbuds (turions) are often essential for proper identification. Winterbuds are formed in late summer and fall as highly condensed clusters of green, apical leaves.

KEY TO SPECIES

1. Leaves (if present) simple, reduced to minute scales (<1 mm long) or bumps, on stiffly upright stems (2–20 cm); floral bracts toothless.......... *M. tenellum*
1. Leaves well-developed, pinnately divided into narrow segments, on elongated flexuous stems; floral bracts toothless, toothed, or deeply divided.......... 2
   2. Leaves alternate, or alternate and whorled on same stem; male flowers with four stamens, 0.5–1 mm long; flowers in axils of submersed leaves or on spikes emersed above water.......... 3
   2. All leaves whorled; male flowers commonly with 8 stamens (4 in *M. heterophyllum*), 1–2 mm long, flowers in emersed spikes.......... 5
3. Flowers and fruits borne in axils of reduced, pinnately divided, bract-like leaves on emersed spikes.......... *M. pinnatum*
3. Flowers and fruits borne in axils of submersed leaves, not emersed above water in a spike.......... 4
   4. Leaves alternate (sometimes near-opposite); fruit 0.7–1.3 mm long, smooth without ridges; no turions present.......... *M. humile*
   4. Leaves both alternate and whorled on same stem; fruit 2–2.5 mm long, warty, with prominent ridges; turions present at stem tips in autumn.......... *M. farwellii*
5. Emersed spikes with bracts (mostly 25–35 mm long, divisions 4–8 mm long) similar to submersed leaves, but rigid; submersed leaves with stalks 5–7 mm long.......... *M. aquaticum*
5. Emersed spikes with conspicuously reduced bracts (<20 mm long, teeth or divisions <4 mm long); submersed leaves with stalks 0–2 mm long.......... 6
   6. Upper flowers of spike and floral bracts alternate; spikes 2–5 cm tall; leaves with 3–7 pairs of narrow segments.......... *M. alterniflorum*
   6. Upper flowers of spike whorled; spikes 4–15+ cm tall; leaves with 5–20 pairs of narrow segments.......... 7

7. Floral bracts inconspicuous, bracts under lower flowers <2 times as long as the flowers or fruits; bracts under upper flowers with toothless to small-toothed edges...... 8

7. Floral bracts conspicuous, bracts under lower flowers ≥2 times as long as the flowers or fruits; bracts under upper flowers with deeply divided to sharp-toothed edges... 9

8. Middle leaves with 5–12 segments on each side; uppermost leaves rounded at tip; cylindrical turions formed in autumn; stems sparingly branched and vertical, stem diameter uniform throughout; non-flowering stem tips usually green and knob-shaped.... ....*M. sibiricum*

8. Middle leaves with 13–20 segments on each side; many uppermost leaves flat-topped at end (as if cut off); turions never formed; stems much-branched, often laying on water surface, and thickened (1.5–2 times as thick as lower stem) just below flowering spikes; non-flowering stem tips usually red and tassel-like......................................*M. spicatum*

9. Bracts under female (i.e., lower) flowers sharply toothed; fruits 1–1.5 mm long; turions not widened at tip, formed at base of stems or rhizomes in fall; stems often reddish.................... .... *M. heterophyllum*

9. Bracts under female flowers pinnately lobed or deeply divided; fruits 2–2.5 mm long; turions club-shaped and widened at tip, formed on stems in fall; stems olive green.............. ....*M. verticillatum*

M. spicatum

M. spicatum

M. sibiricum

ALTERNATE-FLOWERED WATER-MILFOIL *Myriophyllum alterniflorum* DC.

Alternate-flowered water-milfoil is a submersed, densely leaved plant with elongated, flexuous, much-branched stems (0.3–1 mm thick). Leaves are whorled (3–5 per whorl), pinnately divided into 3–7 pairs of narrow segments, and short (3–12 mm long, 3–14 mm wide). Flowers are borne in short (2–5 cm tall) upright spikes, emersed above water, where the upper flowers and floral bracts are alternate (sometimes the lowest are opposite). Floral bracts are short (no more than twice length of flower) and toothless or minutely serrated. Male flowers commonly have 8 stamens (1–2 mm long). Turions never produced.

HABITAT & DISTRIBUTION

Myriophyllum alterniflorum is a northern, native species of mostly shallow, nutrient-poor, lakes and slow streams. It is rare in CT and RI where it is state-listed as ***endangered*** and ***historic***, respectively.

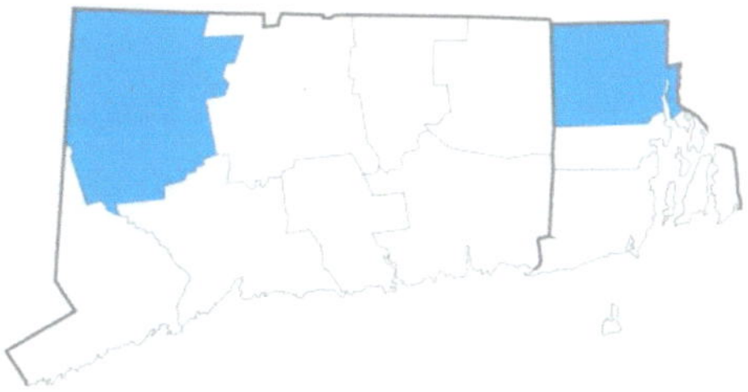

PARROT'S-FEATHER WATER-MILFOIL *Myriophyllum aquaticum* (Vell.) Verdc.

Parrot's-feather water-milfoil is a submersed, densely leaved plant with elongated, flexuous, stems often with considerable leafy portions at tips emersed above water. Submersed leaves are whorled (3–6 per whorl), 8–45 mm long, pinnately divided into 10–15 pairs of narrow segments (4–8 mm long), appearing markedly feather-like, often reddish, with 5–7 mm long stalks. Emersed foliage is crowded, rigid, and grayish-green. Flowers, rarely produced, are borne in axils of emersed, mostly 25–35 mm long, leafy bracts—similar in appearance to submersed leaves. Fruits never produced in our region.

HABITAT & DISTRIBUTION

Myriophyllum aquaticum is an **invasive** species of exposed, shallow, ponds and lakes. It is uncommon in CT and RI (invasive and prohibited in CT). Native to South America, this species was introduced through aquarium and water garden trade. It prefers alkaline nutrient-rich waters and can form thick floating mats.

Notes: The emersed leaves and stems are highly hydrophobic and repel water. Submersed leaves are often shed or brown and deteriorating. Only female plants occur in North America.

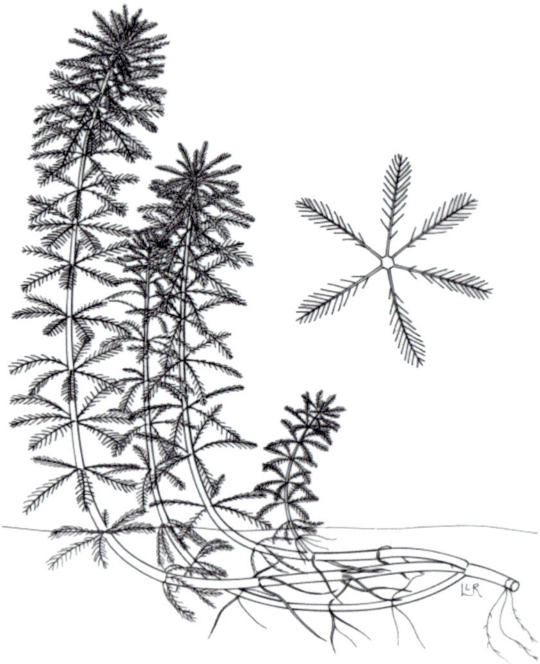

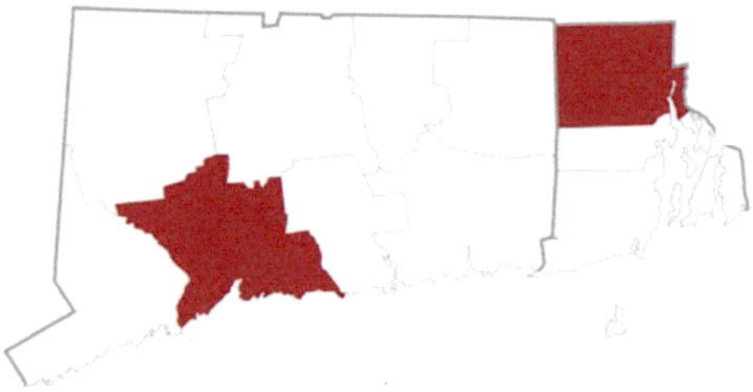

FARWELL'S WATER-MILFOIL *Myriophyllum farwellii* Morong

Farwell's water-milfoil is a delicate, wholly submersed, densely leaved plant with narrow, flexuous stems. Leaves are both alternate and partly whorled on same stem, 10–25 mm long, and pinnately divided into 5–12 pairs of narrow segments. Flowers and fruits submersed, borne in axils of submersed leaves. Male flowers commonly have 4 stamens (0.5–1 mm long). Fruit (2–2.5 mm long) rough, with prominent, bumpy, vertical ridges. Small, stiff turions produced in autumn at the tips of submersed stems.

HABITAT & DISTRIBUTION

Myriophyllum farwellii is a northern native species of cold, acidic, shallow lakes. It is uncommon in CT and RI.

Notes: Very similar to the common and widespread Low water-milfoil (*M. humile*) which produces smooth fruits and lacks turions. The warty fruits of *M. farwellii* are present late July to September.

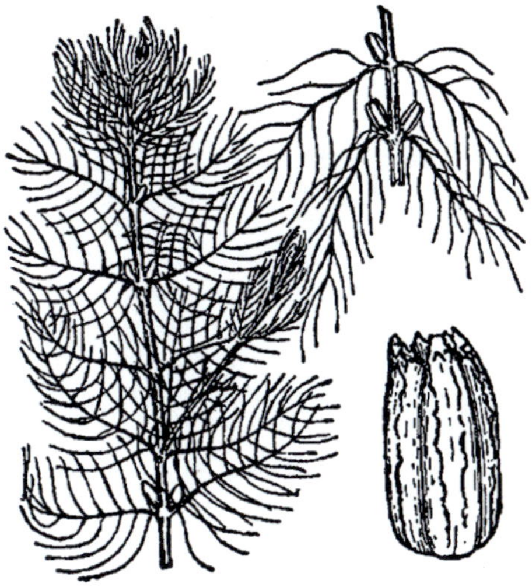

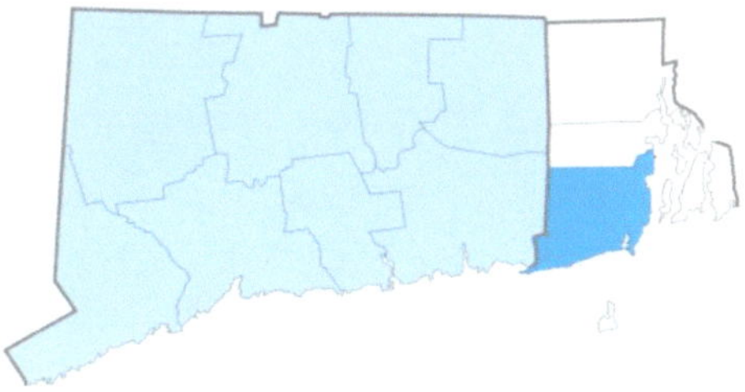

VARIABLE-LEAVED WATER-MILFOIL *Myriophyllum heterophyllum* Michx.

Variable-leaved water-milfoil is a submersed, densely leaved plant with relatively stout, reddish, elongated, much-branched stems, swollen just below the flowering spikes. Submersed leaves (2–4 cm long, 1–3 cm wide) are whorled (4–6 per whorl), closely spaced (whorls <10 mm apart), and pinnately divided into 6–14 pairs of thread-like segments. Flowers are borne in long (15–30 cm tall), leafy, upright, spikes, emersed above water. Floral bracts are whorled, conspicuous, leaf-like, oval (4–18 mm long, much-longer than flowers or fruits), with sharp-toothed edges (bracts near waterline may be somewhat lobed). Male flowers commonly have 4 stamens. Fruits (1–1.5 mm long) are rounded. Turions are produced in fall, at base of stems or rhizomes.

HABITAT & DISTRIBUTION

Myriophyllum heterophyllum is a non-native species of lakes, ponds, and streams. Native to eastern U.S., but this plant is **invasive** and prohibited in New England. It is widespread in CT and RI. Plants can grow rapidly, aggressively spread by stem fragmentation, and develop dense beds which hinder waterbody uses.

In some CT nuisance populations, putative interspecific hybrids (with *M. laxum* Shuttlw. ex Chapman) have been detected.

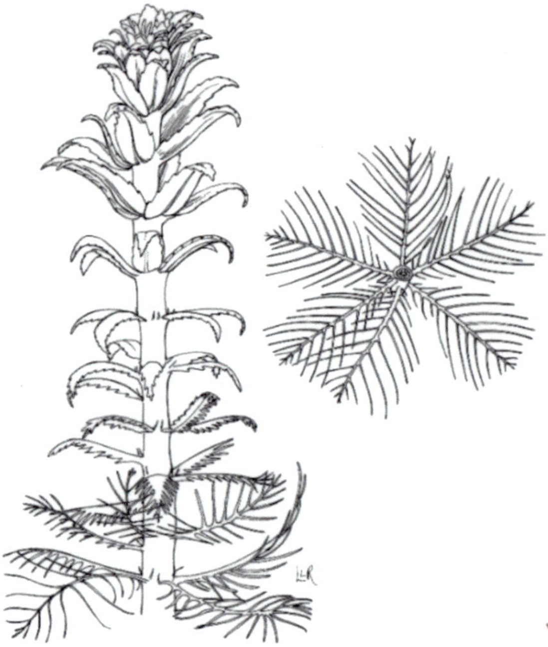

LOW WATER WATER-MILFOIL *Myriophyllum humile* (Raf.) Morong

Low water-milfoil is a delicate, normally submersed, densely leaved plant with elongated, flexuous stems. Leaves are alternate (sometimes some near-opposite), 10–25 mm long, and pinnately divided into 5–12 pairs of narrow segments. Leaves and stems may be brownish red. Flowers and fruits submersed, borne in axils of submersed leaves. Male flowers commonly have 4 stamens (0.5–1 mm long). Fruit (0.7–1.3 mm long) is smooth-sided. No turions produced.

HABITAT & DISTRIBUTION

Myriophyllum humile is a native species in shallow, nutrient poor, mostly acidic, waters of ponds, lakes, and streams. It is widespread and common in CT and RI.

Notes: This northern species is highly variable in form with stem length and leaf number increasing with water depth. When stranded on receding pond shores, a rigid, simple-leaved growth form can occur.

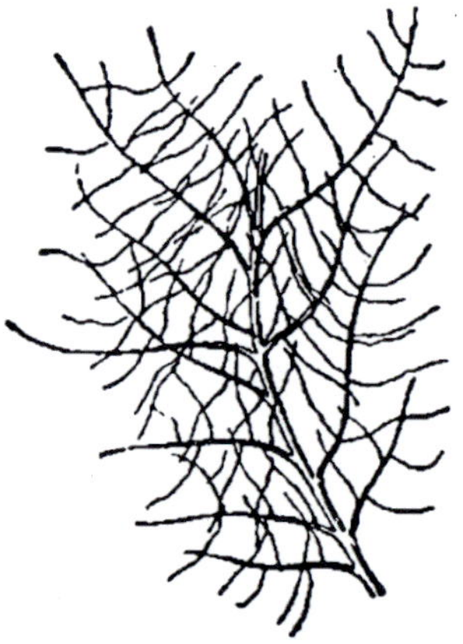

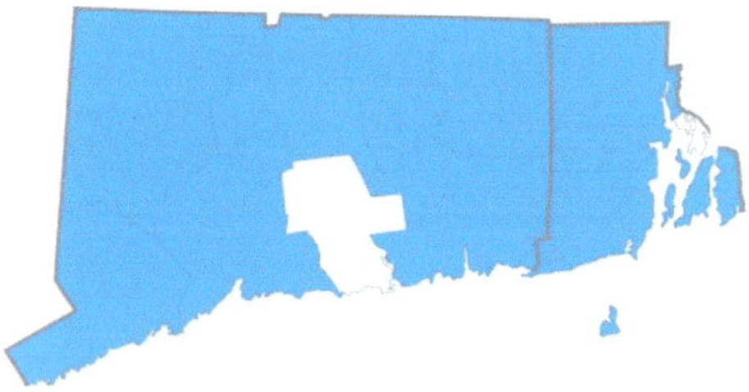

CUT-LEAVED WATER-MILFOIL *Myriophyllum pinnatum* (Walt.) BSP

Cut-leaved water-milfoil is a submersed, densely leaved plant with elongated, slender stems. Leaves (10–30 mm long) are submersed, both alternate and whorled on same stem, and pinnately divided into 8–10 pairs of narrow segments. Flowers are borne on elongated (10–20 cm tall), upright, leafy spikes, emersed above water. Leafy floral bracts are conspicuous (usually more than twice the flower length), alternate, and pinnately divided into flat segments. Male flowers commonly have 4 stamens (0.5–1 mm long). Fruits (1.3–1.8 mm long) have vertical, bumpy ridges on outer surface. Turions never produced.

HABITAT & DISTRIBUTION

Myriophyllum pinnatum is a native, Coastal Plain species occurring in shallow, freshwater or brackish ponds. It is very rare in CT and RI where it is state-listed as ***endangered*** and ***threatened***, respectively.

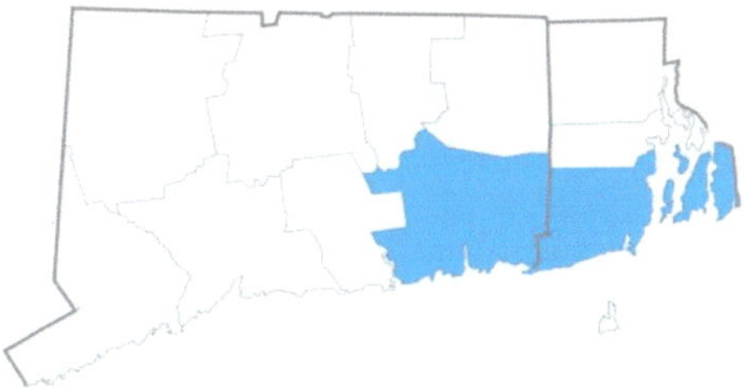

NORTHERN WATER-MILFOIL *Myriophyllum sibiricum* Komarov

Northern water-milfoil is a submersed, densely leaved plant with slender, few-branched, mostly vertical stems. Tips of non-flowering stems are usually green (except during turion formation) and knob-like. Submersed leaves (1.5–3 cm long) are whorled (3–4 per whorl, whorls ≤10 mm apart), and with well-developed middle leaves, pinnately divided into 5–12 pairs of thread-like segments. Many uppermost leaves rounded at tips. Flowers are borne in upright (4–10 cm tall) spikes, emersed above water. Floral bracts are whorled and inconspicuous. Upper bracts are shorter than flowers with toothless to small-toothed edges and lower bracts <2 times as long as the flowers or fruits. Male flowers have 8 stamens. Dark green, cylindrical (≥2 cm long) turions are produced at stem tips in autumn.

HABITAT & DISTRIBUTION

Myriophyllum sibiricum is a native species of primarily clear, alkaline waters. It is very rare in CT, where it is state-listed as ***threatened***.
Note: Similar to Eurasian water-milfoil (*M. spicatum*) but stems essentially unbranched and vertical, fewer leaf segments, and formation of turions in late season.

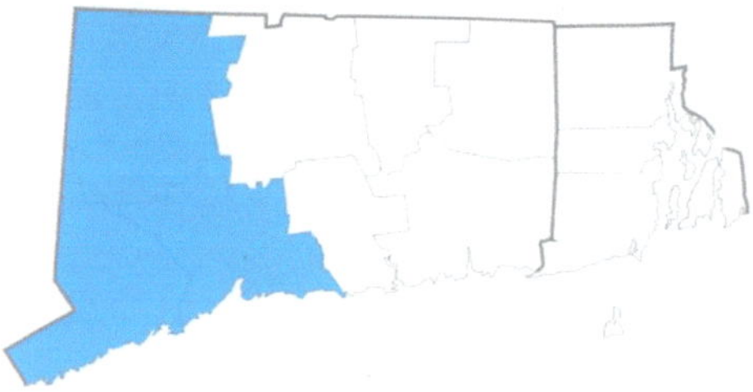

EURASIAN WATER-MILFOIL *Myriophyllum spicatum* L.

Eurasian water-milfoil is a submersed, rhizome-less, densely leaved plant with long (1 m+), slender, much-branched stems, swollen (nearly twice as thick) just below the flowering spikes, and curved to lie parallel to water surface. Tips of non-flowering branches are usually reddish and tassel-like. Submersed leaves (3–3.5 cm long, 1–2.5 cm wide) are whorled (3–4 per whorl), widely spaced (whorls 10–30 mm apart), distinctly stalked, and with well-developed middle leaves, pinnately divided into 13–20 (mostly 14–17) pairs of thread-like segments, and flat (segments all in one plane). Many uppermost leaves flat-topped across at end (as if cut off with scissors). Flowers are borne in upright (6–10 cm tall) spikes, emersed above water. Floral bracts are whorled and inconspicuous. Upper bracts are shorter than flowers with toothless to small-toothed edges and lower bracts <2 times as long as the flowers or fruits. Male flowers have 8 stamens. Turions are never produced.

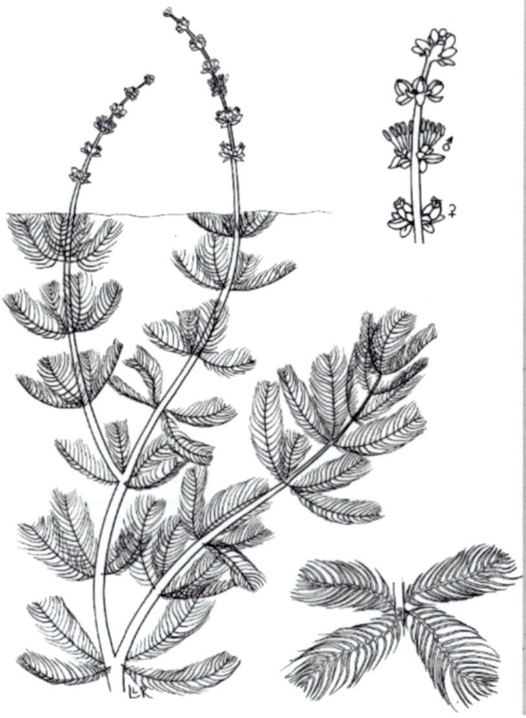

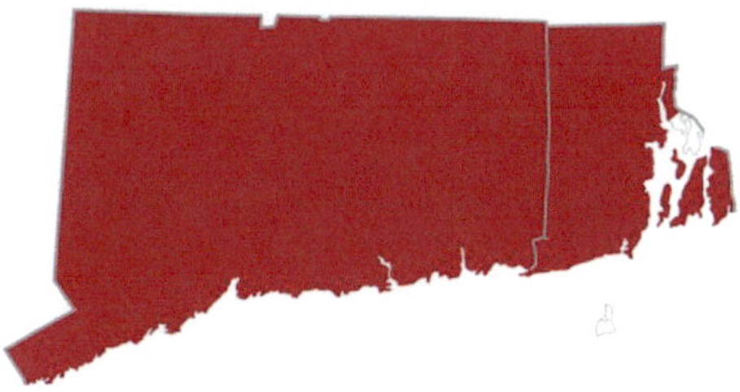

HABITAT & DISTRIBUTION

Myriophyllum spicatum is a non-native species established in lakes and ponds. It is **invasive** and prohibited in New England. Occurrences are scattered in CT and RI.

Notes: This species, a renown aggressive weed, exhibits broad environmental tolerances (e.g., in depth, pH, temperature, salinity, exposure, nutrient load, turbidity), rapid growth, and easy dispersal via stem fragmentation. It is similar to non-flowering plants of Northern water-milfoil (*M. sibiricum*), but has more leaf segments, reddish, tassel-like stem tips, and lacks turions.

Slender water-milfoil *Myriophyllum tenellum* Bigel.

Slender water-milfoil is a slender submersed plant with short (2–20 cm tall), stiff, upright, unbranched stems arising singly from buried rhizomes. Leaves are alternate, inconspicuous, stubby, reduced to minute scales (<1 mm long), and few, or absent. Flowers are borne in the upper leaf axils of wiry, emergent stems. Floral bracts are alternate, toothless, and about the length of the flowers. Fruits are about 1 mm long. Turions never produced.

Habitat & Distribution

Myriophyllum tenellum is a native species of shallow, mostly acidic, ponds and lakes. It is uncommon in Ct and Ri.

Notes: This northern species is unlike other water-milfoils in New England as plants are essentially leafless. Sterile plants are usually abundant, and populations can be dense, propagating largely by vegetative means. Flowering is infrequent and occurs only in very shallow (e.g., 5 cm deep) water where upright stems may reach 30 cm.

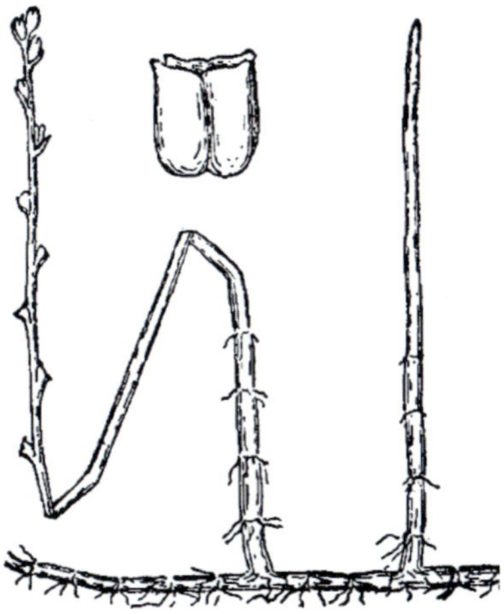

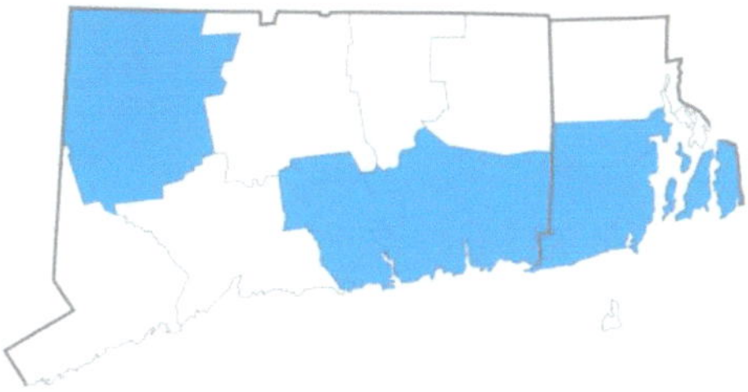

Whorled water-milfoil *Myriophyllum verticillatum* L.

Whorled water-milfoil is a submersed, densely leaved plant with elongated, olive green, sparingly-branched stems. Submersed leaves (2–4.5 cm long, 1.5–2.5 cm wide) are whorled (4–5 per whorl) and pinnately divided into 9-17 pairs of thread-like segments. Flowers are borne in long (7–20 cm tall), leafy, upright, spikes, emersed above water. Floral bracts are whorled, conspicuous (to 20 mm long), lower bracts ≥2 times as long as the flowers or fruits, with mostly deeply divided or pinnately lobed edges. Male flowers commonly have 8 stamens. Fruits (2–2.5 mm long) are rounded. Dark, yellow green, distinctively club-shaped (wider at tip than at base) turions (1–3 cm long) are formed on stem tips in fall.

Habitat & Distribution

Myriophyllum verticillatum is a native species of lakes and streams, with an affinity for alkaline waters. It is uncommon in CT.

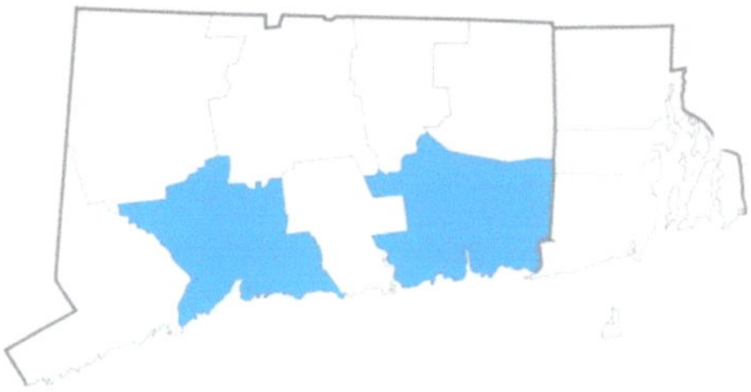

Najas — Hydrocharitaceae

Waternymph

The Waternymphs, or Naiads, are slender, rooted, much-branched, submersed annuals. Their leaves are opposite (or appearing whorled when extra leaves present), toothed along margins, with linear (<2 mm wide) blades that widen to a stem-sheathing base. Flowers are unisexual (on same plant), submersed, and borne 1–3 per leaf axil. Fruits are small (2–4 mm long), with a thin, membranous covering, one-seeded, and tapered at tip.

Four species occur in CT and RI, one being non-native. Fruits and foliage serve as important waterfowl food.

Key to Species

1. Leaf margins with 20–100 minute teeth per side (visible with hand lens); top of basal leaf sheath tapered or sloping.......... 2
1. Leaf margins with 7–17 teeth per side; top of basal leaf sheath distinctly lobed or abruptly widened.......... 3
   2. Leaves tapered from about the middle, to long, slender, pointed tips; seeds smooth and glossy, widest above the middle, deep brown to yellow.......... *N. flexilis*
   2. Leaves abruptly pointed, tapered only from last 2–3 mm, or rounded at tip; seeds minutely pitted and dull, widest at middle, yellowish white with purple tinge, greenish, or reddish brown.......... *N. guadalupensis*
3. Leaves generally stiff and recurved in age, with 7–15 course teeth per side (visible without magnification), top of leaf sheath lobed and finely toothed; seeds slightly curved... *N. minor*
3. Leaves spreading and lax in age, with 13–17 fine teeth per side (visible with magnification), top of leaf sheath abruptly widened, but not lobed, and coarsely jagged; seeds straight.......... *N. gracillima*

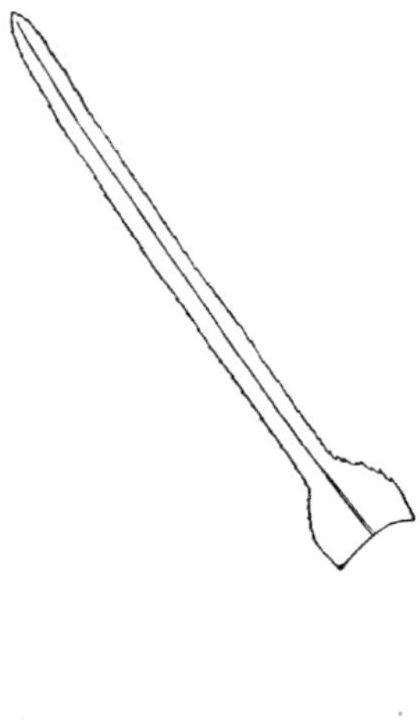

N. flexilis

N. minor

NORTHERN WATERNYMPH *Najas flexilis* (Willd.) Rostk. & Schmidt

Northern waternymph is completely submersed, 25–50 cm tall, with much-branched stems. Leaves are 1.5–4 cm long, 0.5–1 mm wide (above widened base), tapered to a slender tip, with 35–80 fine teeth per side (visible with hand lens). Top of basal leaf sheath is gradually tapered. Usually 1 fruit per leaf axil. Seeds smooth and shining, deep brown to yellow, and widest above the middle.

Habitat & Distribution

Najas flexilis is a native species of lakes, ponds, and streams. It is common and widespread in Ct and Ri.

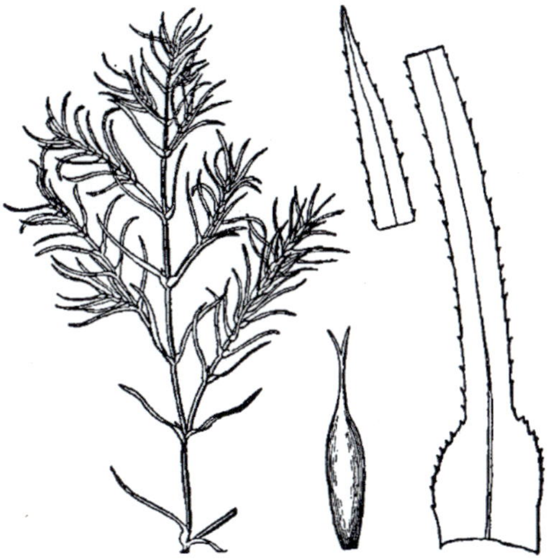

SLENDER WATERNYMPH *Najas gracillima* (A. Braun) Magnus

Slender waternymph is completely submersed, 4–48 cm tall, with slightly branched, slender (to 0.7 mm wide) stems. Leaves are 0.6–2.8 cm long, 0.1–0.5 mm wide (above widened base), pointed at tip, with 13–17 fine teeth per side (visible only with magnification). Top of basal leaf sheath is abruptly widened, but not lobed, and coarsely jagged. Fruits with style positioned off to one side of apex. Seeds straight, dull, light brown.

HABITAT & DISTRIBUTION

Najas gracillima is a native species of lakes and streams, most abundant in acidic waters. It is widespread in CT and RI. Most similar vegetatively to Brittle Waternymph (*N. minor*) however its leaves do not become curved and brittle later in season.

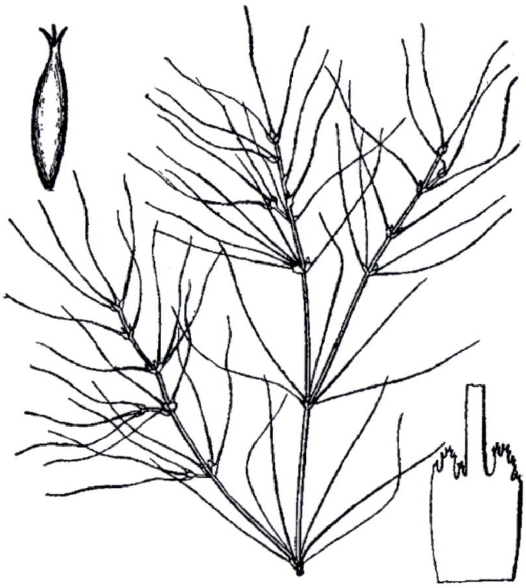

SOUTHERN WATERNYMPH *Najas guadalupensis* (Spreng.) Magnus

Southern waternymph, or Guppy grass, is completely submersed, 10–90 cm tall, with much-branched stems. Leaves are 0.5–3 cm long, 0.2–2.1 mm wide (above widened base), abruptly pointed or rounded at tip, with 20–100 fine teeth per side (visible only with magnification). Top of basal leaf sheath is gradually tapered. Seeds minutely pitted and dull, yellowish white with purple tinge, greenish, or reddish brown, and widest at middle.

HABITAT & DISTRIBUTION

Najas guadalupensis is a native species of alkaline, fresh, or brackish lakes, reservoirs, ponds, and rivers. It is uncommon in CT and RI.

Notes: Two variants of this species exist in CT: finer-stemmed (0.1–0.8 mm wide) plants with 50–100 teeth on each leaf margin represent the more widespread ssp. *guadalupensis*, while those with more robust stems (1–2 mm wide) and 20–40 teeth represent the uncommon ssp. *olivacea* (Rosendahl & Butters) Haynes & Hellquist.

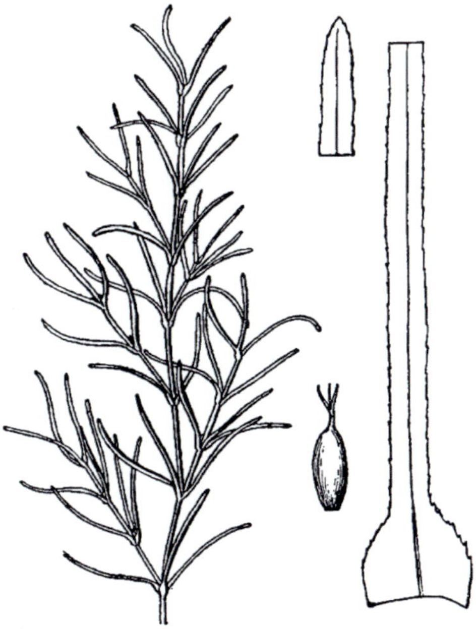

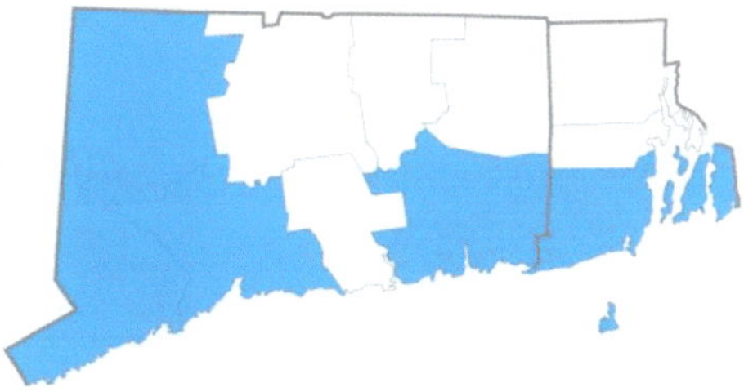

Brittle waternymph *Najas minor* All.

Brittle waternymph, or Brittle naiad, is completely submersed, 11–120 cm tall, compact, with heavily branched stems. Leaves are stiff and recurved in age, 0.5–3.4 cm long, 0.1–1.2 mm wide (above widened base), pointed at tip, with 7–15 course teeth per side (visible without magnification). Top of basal leaf sheath is lobed and finely toothed. Seeds are slightly curved, dull, and purplish brown.

Habitat & Distribution

Najas minor is a non-native species naturalized in lakes and slow rivers. It is widespread in CT and RI where it is **invasive** and banned. Native to Eurasia, the species can be a problematic weed. Plants are easily spread by stem fragmentation and highly tolerant of turbidity and nutrient-rich waters.

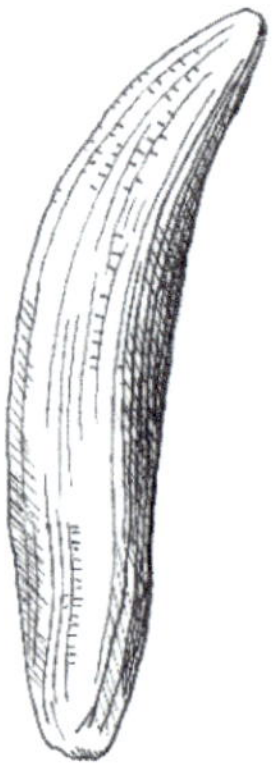

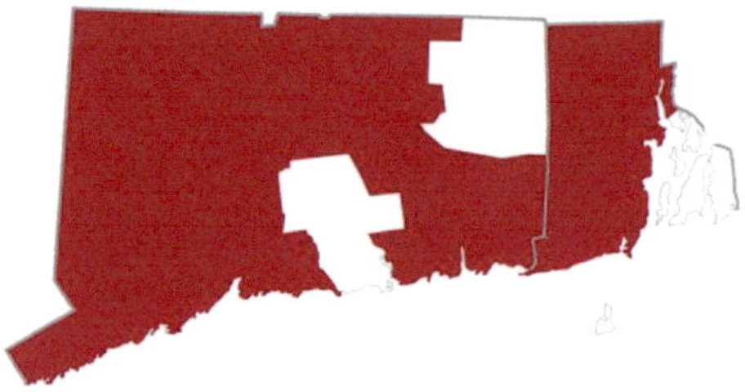

Nelumbo — Nelumbonaceae

Lotus

The Lotuses are robust, rooted, perennials with spongy, buried rhizomes. Their circular, bluish green leaves, either floating or emersed, are usually massive, with stout, centrally attached stalks. The satiny leaf blade surfaces possess an extreme ability to repel water. Flowers are likewise massive (>10 cm wide), showy, held above leaves, and capable of generating (and regulating) heat to attract pollinators. Hard, 1-seeded fruits are embedded in leathery, flat-topped structures.

Two species occur in the region, both non-natives. The species are exceedingly similar to each other, yet very distinguishable from all other aquatic plants. Both are cultivated as ornamental water garden plants with numerous cultivars produced between and within species. Interestingly, their seeds can remain viable for extraordinary periods of time (>150 years).

Key to Species

1. Flowers pale yellow, outermost 1–5 petals usually persistent after others are shed… *N. lutea*
1. Flowers pink, all petals easily detached and shed at an early stage……................. *N. nucifera*

N. lutea

AMERICAN LOTUS *Nelumbo lutea* Willd.

American lotus, or Water lotus, is a rooted, perennial with banana-like rhizomes. Leaves are circular, without a sinus, large (to 60 cm), floating (flat) or emergent (concave), blue green, with centrally radiating veins, and long (to 2 m), centrally attached stalks. Flowers are very large (10–20 cm wide), bisexual, pale yellow, borne singly, above water, on long stalks (to 2 m). Flowers have a central, green, cone-shaped, flat-topped, receptacle. Fruits are nut-like, roundish (10–16 mm long), and embedded in a leathery receptacle (abruptly tapered at base).

HABITAT & DISTRIBUTION

Nelumbo lutea is a non-native species (yet native south of New England) occurring in still, usually shallow (<1 m), waters of lakes, ponds, or tidal marshes, in full sunlight. It is uncommon in CT and RI. **Note:** This plant is fast-growing and can become aggressive.

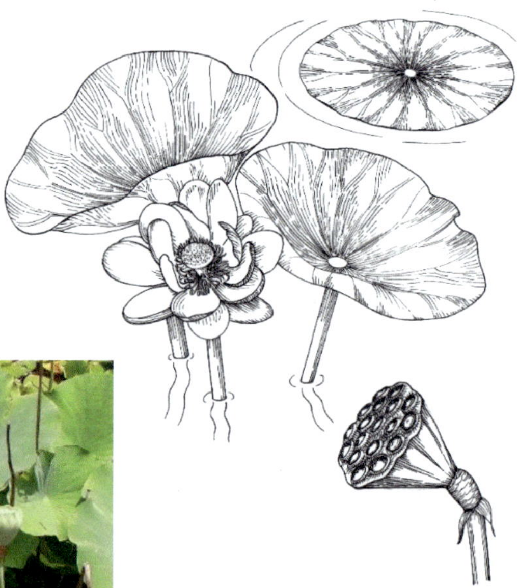

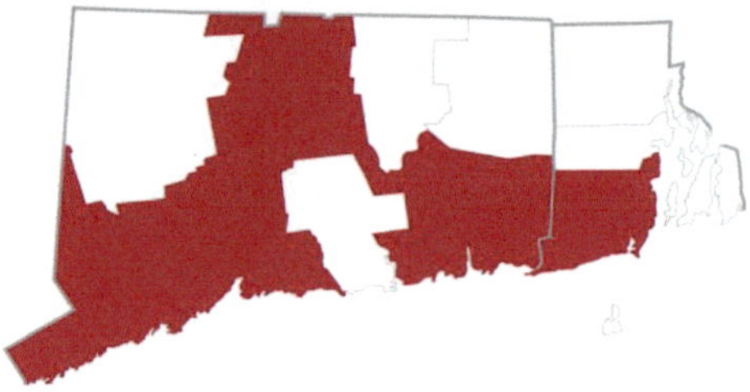

SACRED LOTUS *Nelumbo nucifera* Gaertn.

Sacred lotus, or Indian lotus, is a rooted, perennial with long, slender, horizontal rhizomes. Leaves are circular, without a sinus, large (to 60+ cm wide), floating (flat) or emergent (concave), blue green, with centrally radiating veins, and long (to 2 m), centrally attached stalks. Flowers are very large (10–20 cm wide), bisexual, pink to pinkish-white, fragrant, borne singly, above water, on long stalks (to 2 m). Flowers have a central, green, cone-shaped, flat-topped, receptacle. Fruits are nut-like, oval (10–20 mm long) and embedded in a leathery receptacle (gradually tapering to rounded at its base).

HABITAT & DISTRIBUTION

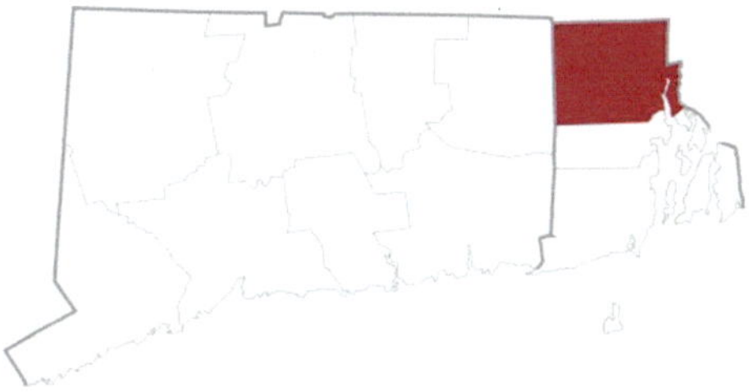

Nelumbo nucifera is native to eastern Asia and sporadically **introduced** as a water garden escape. It occurs in still, mostly acidic, waters of lakes or ponds, with full sun exposure. It is uncommon in RI.

Note: For thousands of years, this plant has been grown as an ornamental and utilized in traditional Asian medicine.

Nuphar — Nymphaeaceae
Pond lily

The Pond lilies, or Spatterdocks, are rooted, perennials with broad, leathery, exposed leaves (mostly floating, sometimes emersed) and submersed membranous leaves. Foliage is generally heart-shaped, with rounded basal lobes, and a defined mid-vein/pinnate venation. Flowers are showy, spherical, and bright yellow, emersed singly atop a stout, rigid stalk. Flowers are bisexual with 5–6 petal-like sepals, inconspicuous scale-like petals, and numerous strap-like stamens surrounding a central, flattened stigmatic disk. Fruits are green to purplish, leathery, and berry-like.

Three species, all native, occur in CT and RI. One species is common and two are rare (and considered extirpated). The name *Nuphar lutea* (L.) Sm., a strictly Eurasian species, is often misapplied to Pond lilies of North America.

KEY TO SPECIES

1. Exposed leaf blades large (13–36 cm long), leaf stalks 4–12 mm wide ; flowers large (2–4.5 cm wide) with 6 sepals and a yellow disk; rhizomes stout (2.5–7 cm diam.)........................2
1. Exposed leaf blades small (4–13 cm long); flowers small (1–2 cm wide) with 5 sepals and a dark red, deeply-lobed disk; rhizomes slender (0.5–2.5 cm diam.) ....*N. microphylla*
   2. Exposed leaves floating, leaf stalks flattened and laterally winged; inside surface of sepals purplish...*N. variegata*
   2. Exposed leaves emersed, sometimes floating, leaf stalks round in cross-section; inside surface of sepals usually green.. *N. advena*

IMMIGRANT POND LILY *Nuphar advena* (Ait.) Ait. f.

Immigrant pond lily is a robust plant with stout rhizomes. Exposed leaves are usually emersed (sometimes floating), large (12–40 cm long) and broad (10–31 cm wide), green, with stout (4–11 mm wide) rounded stalks. Flowers (2–4.5 cm wide) have 6 sepals (inside surface green), long anthers (3–9 mm) and a yellow disk. Fruits are barrel-shaped (1.6–5 cm wide), green, and strongly vertically ribbed.

HABITAT & DISTRIBUTION

Nuphar advena is native to lakes, ponds, and tidal rivers. It is rare in CT where it is state-listed as ***special concern***, considered extirpated.

Notes: This species is common in southeastern US and reaches its northern limits in New England.

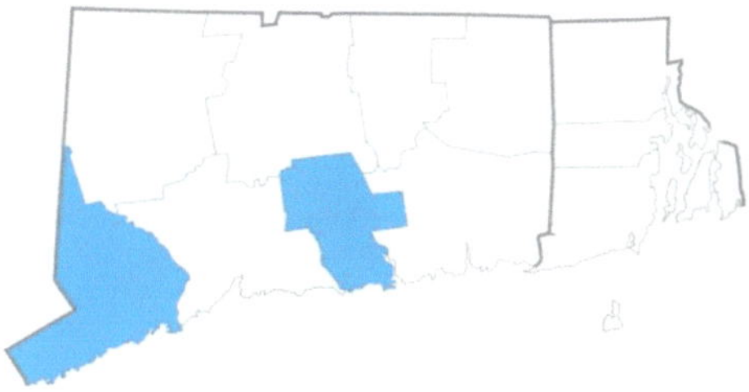

SMALL-LEAVED POND LILY *Nuphar microphylla* (Pers.) Fern.

Small-leaved pond lily, or Tiny cow lily, has floating leaves arising from slender, submersed rhizomes. Floating leaves (4–13 cm long, 3–8 cm wide), green to purple, typically have a deep sinus (up to half the blade length) with weak, compressed stalks (1–2.5 mm wide). Flowers (1–2 cm wide) have 5 sepals, short anthers (1–3 mm) and a dark red, deeply-lobed disk. Fruits are urn-shaped (1–2 cm wide), with an obvious neck and smooth wall.

HABITAT & DISTRIBUTION

Nuphar microphylla is a native species in lakes, ponds, and slow rivers. It is rare in CT where it is state-listed as ***special concern***, but considered historic. This diminutive, quite distinctive, pond lily is imperiled throughout much of its range.

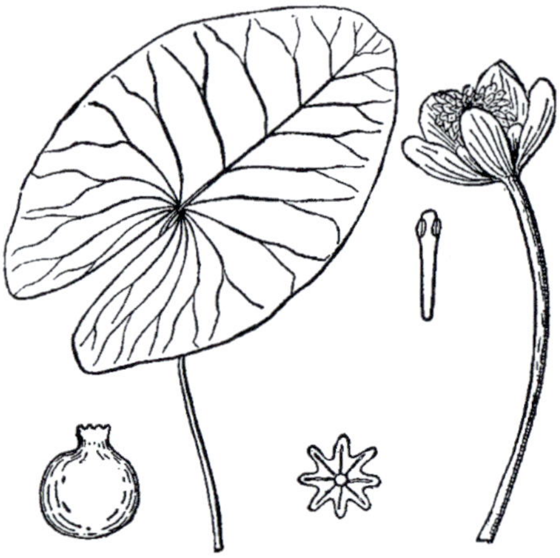

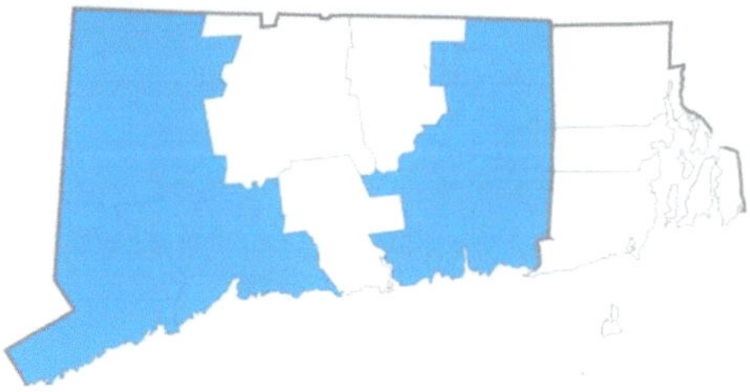

YELLOW POND LILY *Nuphar variegata* Dur.

Yellow pond lily, or Spatterdock, is a is a robust plant with stout rhizomes. Floating leaves are large (13–36 cm long) and broad (10–24 cm wide), green to purplish, with broad (4–10 mm wide), flattened, stalks that usually winged along their length. Flowers (2.5–4.5 cm wide) have 6 sepals (inside surface purple-tinged), long anthers (3–11 mm) and a yellow disk. Fruits are barrel-shaped (1.5–4 cm wide), green to purple, and strongly vertically ribbed.

HABITAT & DISTRIBUTION

Nuphar variegata is native to lakes, ponds, bogs, ditches, and slow-moving rivers.

Notes: It is widespread, common, and usually abundant in CT and RI. This species, usually a conspicuous component of an aquatic flora, is ecologically tolerant, and can colonize deeper (to 2 m) waters. Its long, stout, branched rhizomes can be uprooted by storms and beavers and subsequently float.

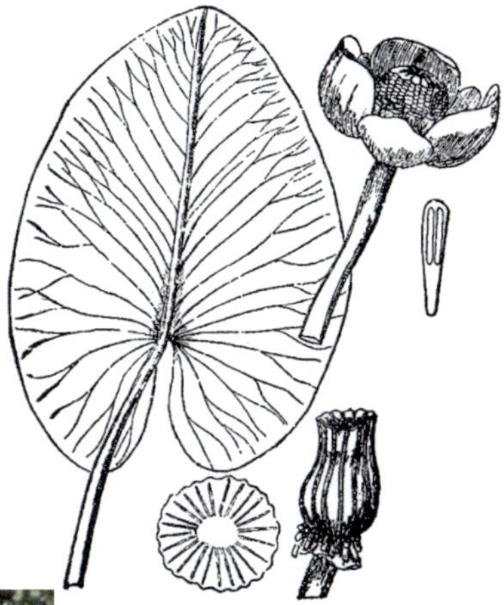

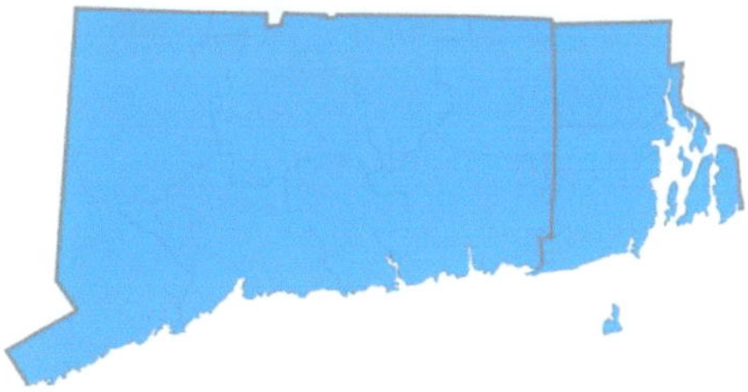

Nymphaea — Nymphaeaceae
Water lily

FRAGRANT WATER LILY *Nymphaea odorata* Ait.

Fragrant water lily is a rooted, perennial, with thick, elongated, horizontal rhizomes. Plants have broad, leathery, floating leaves with long, flexible, green or purple, leaf stalks. Leaves are generally circular in shape (8–40 cm wide), green above and usually purple below, with centrally radiating veins, and a closed sinus with pointed lobes. Flowers are large (5–19 cm wide), showy, fragrant, with yellow centers and numerous (17+), white to pale pink, petals. Flowers are bisexual and emersed just at the water surface (as if floating), borne singly atop stout stalks. Fruits are globose (4 cm), leathery, and berry-like, maturing underwater on coiled stalks.

HABITAT & DISTRIBUTION

Nymphaea odorata is a conspicuous, native species of still or slow waters of lakes, bogs, marshes, swamps, and rivers. It is widespread and common in CT and RI. It has broad tolerances of water depth (0.5–3 m), nutrients, and pH levels. Flowers are insect-pollinated and exhibit diurnal blooming (opening in early morning and closing late afternoon) over a 3-day period.

Notes: Two subspecies are represented in CT. Unlike typical plants, ssp. *tuberosa* (Paine) Wiersema & Hellquist is more robust with rhizomes forming numerous, detachable tubers. Its leaves have green leaf stalks with purplish stripes and blades mostly green underneath. This uncommon variant is considered introduced into CT.

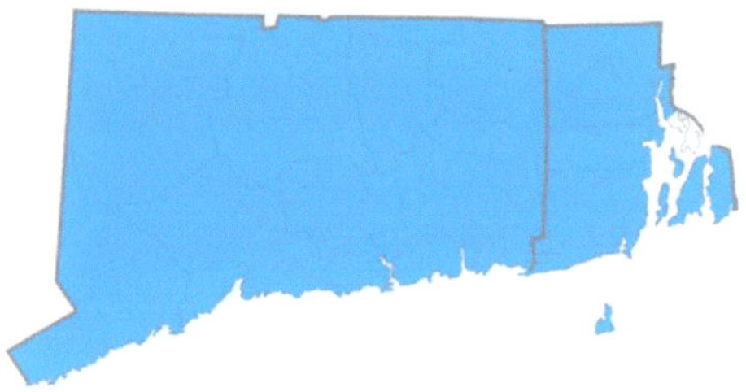

ssp. *odorata*

ssp. *tuberosa*

Nymphoides — Menyanthaceae
Floating-heart

The Floating-hearts are rooted, perennials of quiet waters with shiny, heart-shaped floating leaves which superficially resemble those of juvenile waterlilies (*Nuphar* or *Nymphaea*). Plants have slender, anchored rhizomes and long (to 2 m), buoyant, flexuous stems that bear leaves and flower clusters at their summit. Flowers are showy, with 5 united petals, emersed or floating, and borne in clusters from the base of the leafstalk—appearing as if borne on a long leaf stalk. Fruits are elliptical, green capsules.

Two species occur in CT and RI, one is non-native. Both species are highly ornamental and cultivated for the water garden trade.

KEY TO SPECIES

1. Flowers white, 5–10 mm wide; flowering stems with a single leaf (3–5 cm long) at base of flower cluster; clusters of submersed spur-like roots present at the leaf base, near water surface.. *N. cordata*
1. Flowers yellow, 20–25 mm wide; flowering stems with a pair of leaves (each 5–15 cm long) at base of flower cluster; usually lacking spur-like roots at base of leaves........... *N. peltata*

LITTLE FLOATING-HEART *Nymphoides cordata* (Ell.) Fern.

Little floating-heart is a submersed plant with floating leaves produced singly from the base of a flower cluster. The floating blades are heart-shaped (3–5 cm long), with rounded lobes, and usually mottled with purple. Submersed clusters of slender (2–4 mm thick), fleshy roots are produced at the leaf base, near water surface. Flowers (5–10 mm wide) are white with yellow centers, bloom above water, with 5 fused petals. Fruits are small (4–5 mm long) capsules.

HABITAT & DISTRIBUTION

Nymphoides cordata is a native species in lakes, ponds, and slow streams. It is widespread in CT and RI.

Notes: Plants prefer shallow waters (<1 m) and are tolerant of either nutrient-poor or -rich systems. The clusters of tuber-like roots with attached floating leaves can detach as plantlets for dispersal purposes.

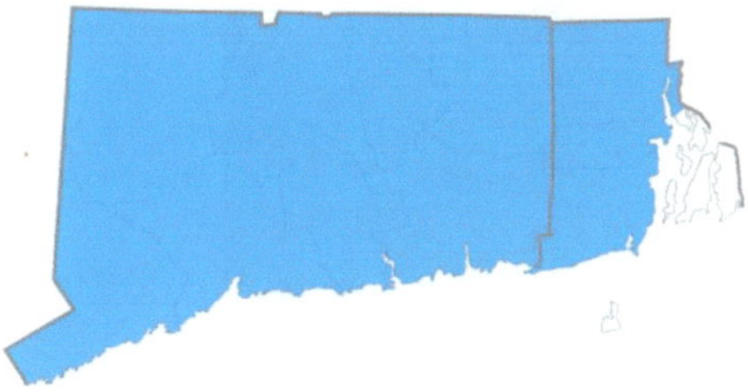

YELLOW FLOATING-HEART *Nymphoides peltata* (Gmel.) Ktze.

Yellow floating-heart is a submersed plant with floating leaves produced in pairs from the base of a flower cluster. The floating blades are heart-shaped (5–15 cm long), with rounded lobes, and wavy margins. Flowers (20–25 mm wide) are wholly bright yellow, bloom above water, with 5 fused petals. Fruits (12–25 mm long) are beaked, flattened, capsules containing seeds edged with stiff hairs.

HABITAT & DISTRIBUTION

Nymphoides peltata is a non-native species of lakes, ponds, and slow rivers. It is uncommon in CT and RI. This Eurasian native, introduced for ornamental purposes, is **invasive** and prohibited in New England. **Notes:** Plants are generally stout and highly productive, most common in nutrient-rich waters, and tolerant of waters to 4 m deep.

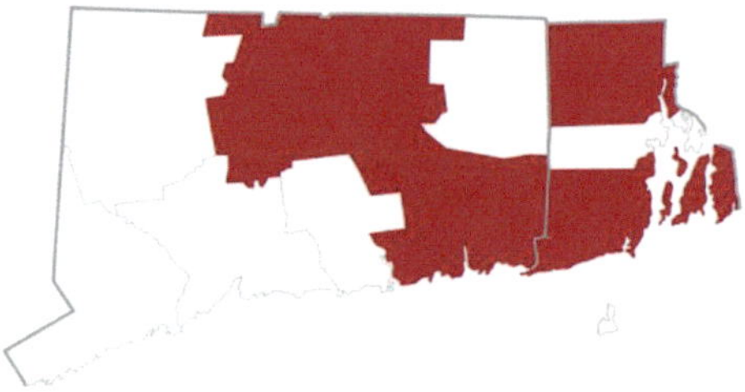

Orontium — Araceae
Golden Club

GOLDEN CLUB *Orontium aquaticum* L.

Golden Club is a highly distinctive perennial hydrophyte. Plants have a thick, rooted rhizome, and erect or floating leaves. Leaves are blue green, borne in a compact, basal cluster, with stout, long (10–40 cm), fleshy stalks and somewhat fleshy, oblong to elliptical, pointed, parallel-veined blades (10–30 cm). Flowers are minute, bright yellow, and borne in dense clusters at the end of long (2–10 cm long), thick (5–10 mm wide), white stalks, extending above water surface. Fruits are berry-like, slimy, spherical (6–10 mm), dark green, and 1-seeded.

HABITAT & DISTRIBUTION

Orontium aquaticum is a native species of shallow waters (to 1 m) of ponds, freshwater tidal marshes, and shores of streams. This Coastal Plain species reaches the northern end of its range in CT and RI where it is rare and state-listed as ***special concern*** and ***endangered***, respectively.

Notes: Its velvety & waxy leaf surfaces are resistant to wetting (hydrophobic). It is spring-flowering and produces buoyant fruits from which seeds can germinate while floating. It is cultivated as a water garden ornamental.

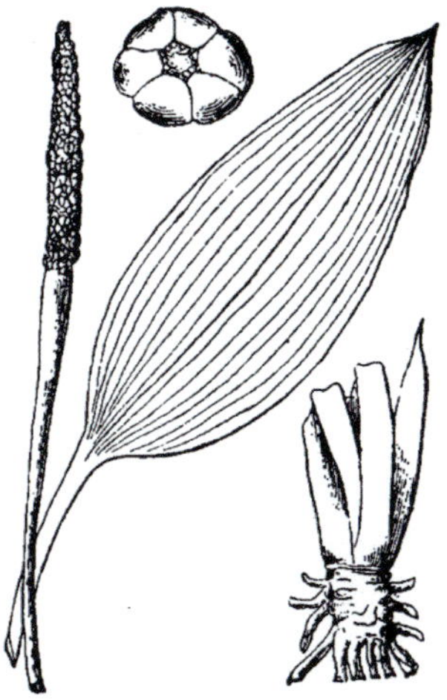

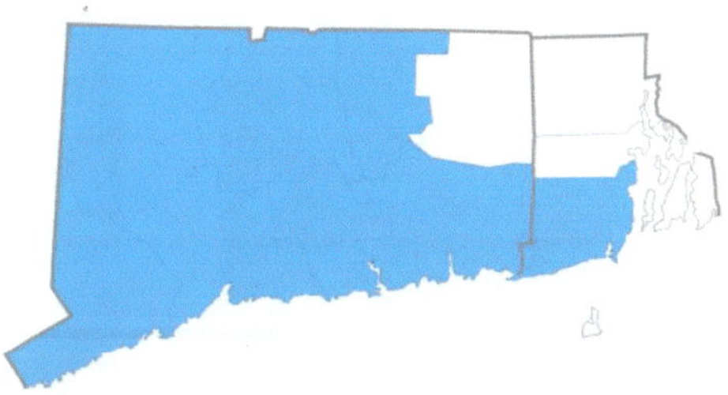

Persicaria – Polygonaceae
Smartweed

WATER SMARTWEED *Persicaria amphibia* (L.) Gray

Water smartweed is a rooted, floating-leaved perennial with rhizomes that grow prostrate, and sprawling, branched stems. Leaves are alternate, floating, hairy to hairless, short-stalked, with elliptical blades (2–15 cm long, to 17–29 mm wide) and pointed or rounded tips. There is a cylindrical stipule (an ocrea) sheathing the stem at each node. Flowers are small (4–6 mm long), pink, 5-lobed, bisexual, and borne in an erect, dense, many-flowered spike-like cluster (10–30 mm long) that terminates the stem above water. Fruits are small (2–3 mm), flattened, dark brown, shiny, achenes.

HABITAT & DISTRIBUTION

Persicaria amphibia is native species of shallow, still or slow-moving waters in marshes, lakes, and rivers. It is uncommon in CT and RI.

Notes: A very variable species, typically growing underwater or floating, but even emersed if water levels drop. It is similar to the more widespread emergent wetland species Scarlet smartweed [*P. coccinea* (Muhl. ex Willd.) Greene] which can grow floating leaves if it becomes inundated. It has much wider leaves (32–63 mm) and longer (30–90 mm) flower clusters.

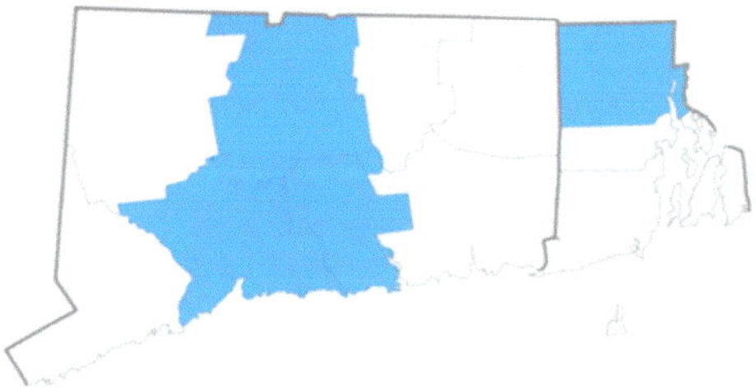

Pistia — Araceae
Water Lettuce

WATER LETTUCE *Pistia stratiotes* L.

Water lettuce, or Tropical duckweed, is a free-floating perennial with tufted leaves, long (to 50 cm) feathery roots hanging in water column, and short horizontal stems connecting plants. Leaves (2–15 cm long) are in a dense basal rosette, floating to emersed, light green, velvety hairy, parallel-veined, and spongy. Flowers, if present, enclosed in an erect, hairy, greenish, cluster at center of plant.

HABITAT & DISTRIBUTION

Pistia stratiotes is considered native to warmer regions of southern U.S. but introduced into our area in still or slow-moving waters of lakes and rivers. It has been sporadically introduced in CT and RI presumably as a water garden escape.

Notes: This plant can be aggressive and potentially problematic, but it is not known to over-winter in New England.

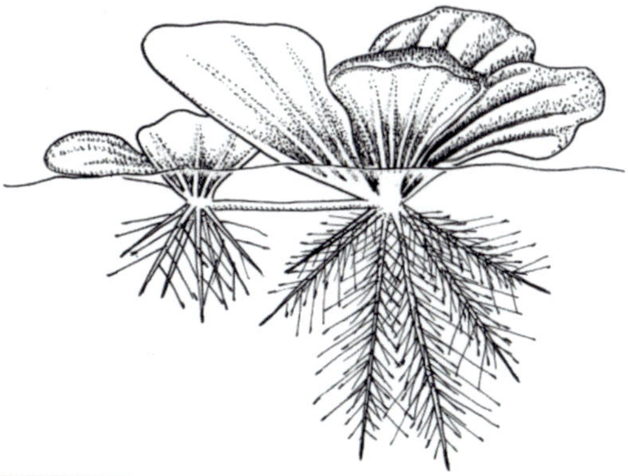

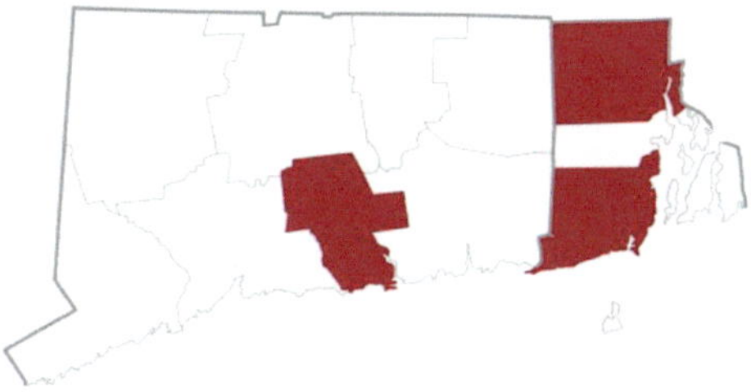

Podostemum — Podostemaceae
Riverweed

HORN-LEAVED RIVERWEED *Podostemum ceratophyllum* Michx.

Horn-leaved riverweed, or Threadfoot, is an olive-green, algal-like, submersed, perennial with flattened, leathery stems, and creeping, green roots adhering to hard substrates via fleshy disks. Leaves are alternate, long (2–142 mm), and repeatedly forked into linear or hair-like (0.05–0.8 mm wide) segments. Flowers are inconspicuous, water-pollinated, bisexual, and borne singly along stems. Fruits are small (1.5–3 mm long), 2-valved capsules.

HABITAT & DISTRIBUTION

Podostemum ceratophyllum is a native species of rocky, fast-flowing rivers and streams of good water quality. It is uncommon in CT and of ***concern*** in RI, but possibly overlooked. Usually growing in areas of rapids, attached to ledge, rocks, or boulders.

Note: This is a highly unusual flowering plant—often mistaken for mosses or algae—occupying a highly specialized habitat. Plants become reproductive as water levels drop and expose them to air.

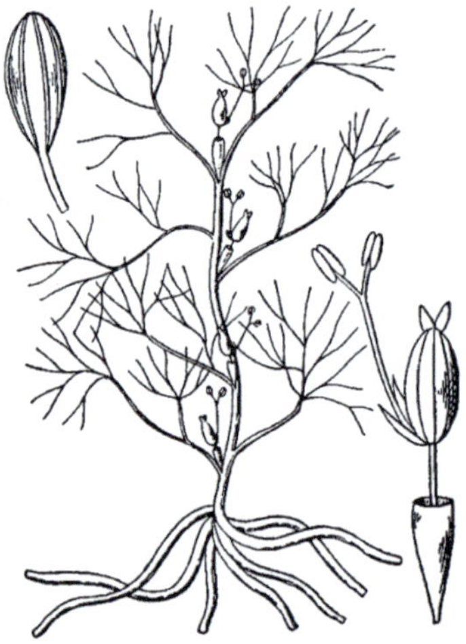

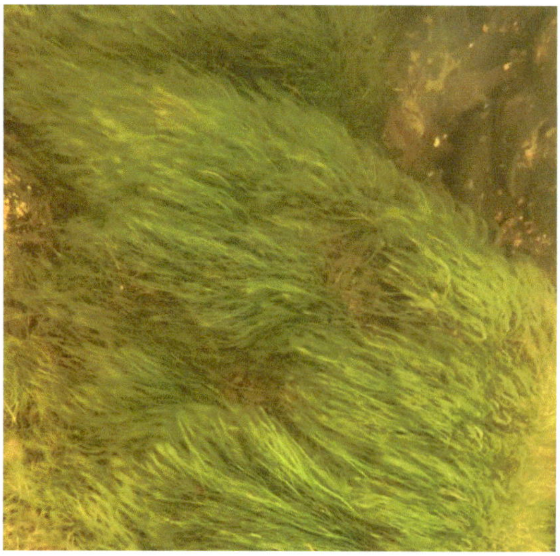

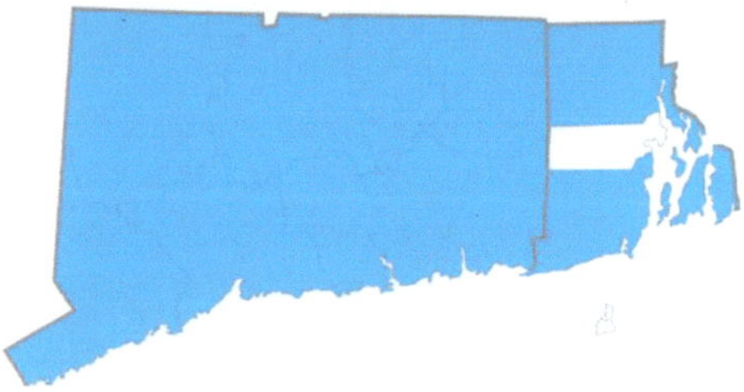

Pontederia – Pontederiaceae
Pickerelweed

PICKERELWEED *Pontederia cordata* L.

Pickerelweed is a submersed (or emersed) perennial with rhizomes rooted in mud. Leaves are submersed in a basal rosette. Blades are without stalks, linear, thin, ribbon-like (6–25 cm long, 4–6 mm wide), and with an observable midvein. Plants can sometimes have narrow emersed leaves.

HABITAT & DISTRIBUTION

Pontederia cordata is a native species of lake shores, ditches, and brackish tidal marshes. It is common throughout CT and RI. Plants are very variable in shape of leaves.

Notes: The submersed life-form (f. *taenia* Fassett) of the species is described here. Typically, the species exists as the larger, emergent, life-form. The emergent form, very common on muddy shores, is more robust overall and has stout, long-stalked (to 60 cm), glossy, heart-shaped leaf blades (to 12 cm wide) and dense spikes of showy, blue-purple flowers borne on upright stems. The submersed life-form is similar to Bur-reed (*Sparganium*) and Arrowhead (*Sagittaria*).

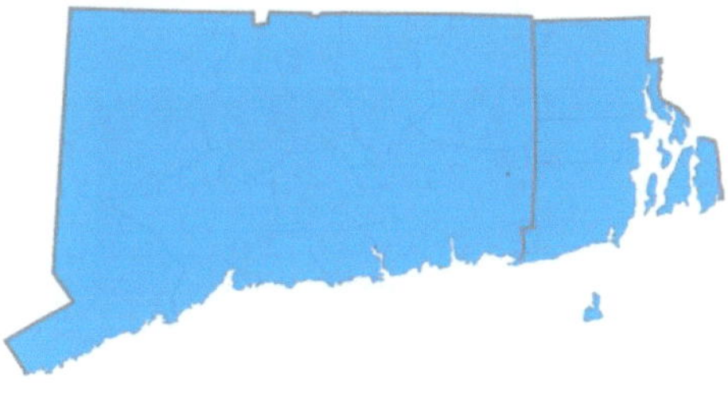

Potamogeton — Potamogetonaceae
Pondweed

The Pondweeds are mostly submersed, rooted perennials with simple or branched stems, regularly with slender, creeping rhizomes. Leaves are alternate (uppermost may appear opposite), mid-veined, and submersed and or floating. All species have submersed leaves, which are thin, ribbon-like or broader, and usually flaccid. Some species additionally have floating blades which are firmer, and elliptical in shape. At the base of each leaf, there is an open (margins unfused) or closed (fused) stipule that forms a cylindrical sheath around the stem. Sometimes a portion of the stipule is fused to base of leaf. Flowers are small, individually inconspicuous, and clustered mostly into narrow, cylindrical, wind-pollinated spikes held above water (or in submersed, few-flowered clusters). Fruits are small, hard, seed-like, achenes. Winter-buds (turions) are produced by some species at tips of branches or in leaf axils. Winter buds are firm, much-condensed stems with crowded stipules and reduced, thickened leaves.

Twenty-six species occur in CT and RI, including six rare, and one introduced species. Pondweeds comprise the most diverse group of aquatic plants in our region and collectively have high ecological importance as wildlife food. Identification is sometimes difficult and may require mature fruits and magnification. Vegetative features are highly variable and influenced by environmental conditions (water depth and flow).

Pondweeds have a propensity for natural interspecific hybridization which can confound identification. Research supports plants long referred to as Ogden's pondweed (*P. ogdenii* Hellq. & Hilton), currently state-listed (endangered) from waters of Litchfield Co., CT, are natural hybrids. Like other known *Potamogeton* hybrids in our area, details of these plants are excluded here.

P. gramineus

P. natans

P. zosteriformis

P. obtusifolius

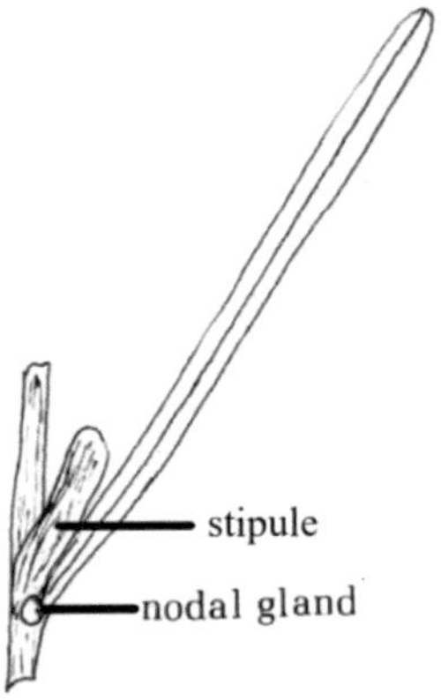

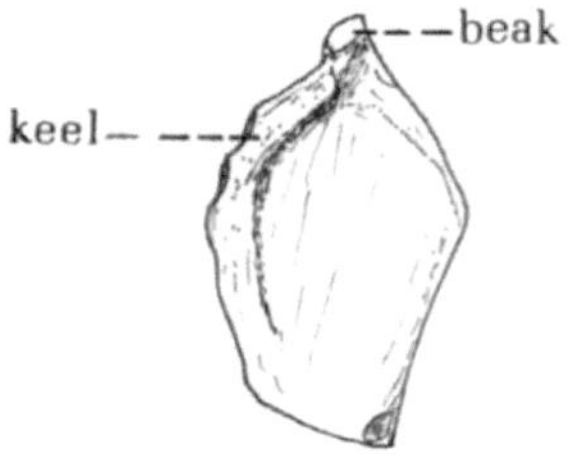

KEY TO SPECIES

1. Stipule fused to base of submersed leaves for a portion its length, remaining tip of stipule extending as a free flap of tissue (ligule)………………………………………………… 2
1. Stipule completely free from leaf base……………………………………………………4
   2. Leaves conspicuously two-ranked so that whole plant is flat; leaf blades stiff, with finely toothed edges, 20–60 veins, and a pair of round lobes at base; plants without floating leaves………………………………………………………………………… *P. robbinsii*
   2. Leaves not conspicuously two-ranked; blades lax, with smooth edges, <20 veins, and not lobed at base; plants with or without floating leaves………………………………………3
3. Submersed leaf blades 0.5–2 mm wide, blunt-tipped; floating leaf tips rounded; fruit ca. 1.5–2.5 mm in diameter (including keel), keel blunt and poorly developed, beak absent………… ……………………………………………………………………………………… *P. spirillus*
3. Submersed leaf blades 0.1–0.4 mm wide, pointed or long-tapering at tip; floating leaf tips acute; fruit ca. 1–1.5 mm in diameter, keel well-developed, beak present or absent………… ……………………………………………………………………………………*P. bicupulatus*
   4. Plants with floating leaves………………………………………………………………… 5
   4. Plants without floating leaves………………………………………………………………13
5. Floating leaf blades 6–15 mm long, 5–9 veined; submersed leaves 0.1–1 mm wide, 1–3 veined; flower spikes 3–8 mm long; fruit 1.5–2.5 mm wide……………………………… *P. vaseyi*
5. Floating leaf blades 15–120 mm long, many-veined (7–49); submersed leaves 0.25–10 mm wide, 1–37 veined; flower spikes 10–50 mm long; fruit 2.0–4.5 mm wide…………………………6
   6. Submersed leaves linear, flattened or phyllodial (appearing as a bladeless leafstalk) …… 7
   6. Submersed leaves broadly linear-oblong, lance-shaped to elliptical………………………… 9
7. Submersed leaves 2–10 mm wide, thin and transparent, 3–13 veined, with prominent lacunae band 1–2 mm wide on each side of mid-vein; fruits with distinct, acute keels on back and sides…………………………………………………………………………*P. epihydrus*
7. Submersed leaves 0.25–2.5 mm wide, thick and nearly opaque, phyllodial (appearing as a bladeless leaf stalk), 3–5 veined, lacking lacunae bands; fruit without keels on sides……… 8
   8. Floating blades 5–10 cm long, usually rounded or somewhat lobed at base, with 17–35 veins, leaf stalk pale and thin at blade end; stipules 4–10 cm long; fruit 3.5–5 mm wide, wrinkled on sides, keel inconspicuous……………………………………………… *P. natans*
   8. Floating blades 2.5–6 cm long, rounded to tapering at base, with 7–23 veins, leaf stalk usually not pale at blade end; stipules 2.5–4 cm long; fruit 2–3.5 mm wide, not wrinkled on sides, with a prominent keel…………………………………………………*P. oakesianus*

9. Submersed leaves without a distinct stalk (petiole)............................... *P. gramineus*
9. Submersed leaves with a narrow stalk 10
10. Floating leaves 19–49 veined; submersed leaf margins undulating (crisped); stems rusty spotted to black-spotted; fruits keeled..11
10. Floating leaves 11–29 veined; submersed leaf margins flat, rarely crisped; stem lacking spots...12
11. Stems black-spotted; submersed leaves with 7–19 veins, 10–25 mm wide, slightly arching; floating leaf blades with 15–19 veins; fruit 3.1–4.1 mm................................. *P. pulcher*
11. Stems rusty-spotted; submersed leaves with 19–49 veins, 25–75 mm wide, distinctly arching; floating leaf blades with 27–49 veins; fruit 3.9–5.2 mm..................... *P. amplifolius*
12. Submersed leaves with long stalks (5–15 cm long), blades with 2–5 rows of lacunae on each side of mid-vein, larger leaves with blades 10–35 mm wide; leaf tips pointed; fruit red to red-brown, the lateral keels with blunt to sharp teeth .... *P. nodosus*
12. Submersed leaves stalkless or with short stalks (0.5–4 cm long), larger leaves with blades 20–45 mm wide; leaf tip with an abrupt, short, stiff point; fruit gray-green to olive-green, the lateral keels without teeth .... *P. illinoensis*
13. Leaf margins finely but distinctly and sharply toothed especially towards tip; fruit with prominent (2–3 mm long), curved, beak; turions commonly formed.....................*P. crispus*
13. Leaf margins without distinct teeth; fruit beak <0.9 mm long; turions absent or rarely formed 14
14. Leaves linear, ribbon-like, thread-like, flattened or phyllodial (appearing as a bladeless leafstalk) 15
14. Leaves broadly linear-oblong, lance-shaped, or elliptical...29
15. Plants with obvious, elongated rhizomes..16
15. Plants lacking rhizomes or rhizomes not apparent...19
16. Leaves delicate, thread-like, 0.1–0.5 mm wide, 1-veined; flowering stalks 5–25 cm long.... ....*P. confervoides*
16. Leaves flattened and ribbon-like or appearing as bladeless stalks, 2–11 mm wide, 3–13-veined; flowering stalks 1.5–8 cm long.. 17
17. Submersed leaves 2–10 mm wide, thin and transparent, 3–13 veined, with prominent lacunae band 1–2 mm wide on each side of mid-vein; fruits with distinct, acute keels on back and sides... *P. epihydrus*
17. Submersed leaves 0.25–2 mm wide, thick and nearly opaque, phyllodial (appearing as a bladeless leaf stalk), 3–5 veined, lacking lacunae bands; fruit without keels on sides......... 18
18. Submersed leaves 0.7–2.5 mm wide, 3–5 veins; fruit 3.5–5 mm wide, wrinkled on sides, keel inconspicuous.. *P. natans*
18. Submersed leaves 0.25–1 mm wide, 3 veins; fruit 2–3.5 mm wide, not wrinkled on sides, with a prominent keel...*P. oakesianus*
19. Leaves 1-veined, 0.1–1 mm wide...20
19. Leaves 3–35 veined, 1.2–5 mm wide...21
20. Circular raised glands present on at least some nodes, green, gold, brown, or rarely white, to 0.5 mm wide; stipules 5–20 mm long; flowering stalks 0.5–6 cm long............ .... *P. gemmiparus*
20. Nodal glands absent; stipules 4–12 mm long; flowering stalks 0.5–3 cm long...... *P. vaseyi*
21. Circular raised glands present on at least some nodes..22
21. Nodal glands absent..27
22. Stipules fibrous, often whitish or brown...23
22. Stipules delicate, not fibrous, green, brown, or white...25

23. Leaf blades obtuse or rounded at tip, sometimes with a short, projected tip; leaves 5–7 veined; winter buds with inner leaves oriented at a right angle to the outer leaves, outer leaves corrugated at the base, the inner leaves modified into a fan-shaped structure………. ………………………………………………………………………………………. *P. friesii*

23. Leaf blades obtuse to tapering at tip; winter buds flattened, the inner and outer leaves oriented in the same plane, the outer leaves not corrugated at the base, the inner leaves undifferentiated……………………………………………………………………………… 24

24. Stems conspicuously flattened, stiff, 0.6–3.2 mm wide; leaf blades with 15–35 veins; fruit 4–4.5 mm wide; stipules conspicuous, 15–35 mm long, with overlapping margins…………. ……………………………………………………………………………………. *P. zosteriformis*

24. Stems not flattened, flexuous, <1 mm wide; leaf blades with 3–5 veins; fruit 1.9–2.1 mm wide; stipules inconspicuous, 6–16 mm long, with margins fused to form a cylinder or tube around stem ………………………………………………………………………… *P. strictifolius*

25. Leaf blades 1–3.5 mm wide, bluntly pointed or rounded at tip, often reddish; fruit 2.5–3.5 mm wide; winter buds 3.5–7.8 cm long, 2.3–5.1 mm wide………………………. *P. obtusifolius*

25. Leaf blades 0.2–2.5 mm wide, pointed at tip, green; fruit 1.5–2.2 mm wide; winter buds 0.9–3.2 cm long, 0.3–1.8 mm wide ……………………………………………………………….26

26. Stipules with fused margins, forming a tube surrounding the stem; leaf blades with up to 2 rows of lacunae on each side of the midvein; flowering stalks 10–62 mm long, usually terminal, 1–3 per plant; spike usually of 2–4 distinct, interrupted whorls of flowers/fruits… ………………………………………………………………………………………. *P. pusillus*

26. Stipules wrapped around the stem, but with unfused, overlapping margins; leaf blades with up to 5 rows of lacunae on each side of the midvein; flowering stalks 5–30 mm long, axillary or terminal, more than 3 per plant; spike usually of 1–3 crowded whorls of flowers/fruits……………………………………………………………………………. *P. berchtoldii*

27. Stems conspicuously flattened (to 3.2 mm wide); leaf blades with 15–35 veins; spikes cylindrical with 7–11 whorls of flower; flowering stems stout and long (20–50 mm); fruits 4–4.5 mm wide………………………………………………………………………. *P. zosteriformis*

27. Stems rounded or slightly compressed (to 1 mm wide); leaf blades with 3–5 veins; spikes knob-like with 1–5 whorls of flowers; flowering stems short (3–14 mm); fruits 1.4–4 mm wide ………………………………………………………………………………………………28

28. Leaves acute at tip; fruit 1.4–2.3 mm wide, with a single, wing-like, bumpy keel; flowering stalks 3–11 mm long and axillary ………………………………………………………… *P. foliosus*

28. Leaves bristle-tipped (rarely blunt); fruit 2.3–4 mm wide, with three keels; flowering stalks 6–14 mm long, axillary and terminal…………………………………………………. *P. hillii*

29. Submersed leaf blades with a stalk……………………………………………………………30

29. Submersed leaves without a stalk………………………………………………………………33

30. Submersed leaf margins undulating (crisped); stems rusty-spotted to black-spotted….31

30. Submersed leaf margins flat, rarely crisped; stems lacking spots………………………………32

31. Stems black-spotted; submersed leaves with 7–19 veins, 10–25 mm wide, slightly arching; fruit 3.1–4.1 mm………………………………………………………………………. *P. pulcher*

31. Stems rusty spotted; submersed leaves with 19–49 veins, 25–75 mm wide, distinctly arching; fruit 3.9–5.2 mm…………………………………………………………. *P. amplifolius*

32. Submersed leaves with long stalks (5–15 cm long), larger leaves with blades 10–35 mm wide; leaf tips pointed; fruit red to red-brown, the lateral keels with blunt to sharp teeth… ……………………………………………………………………………………… *P. nodosus*

32. Submersed leaves stalkless or with short stalks (0.5–4 cm long), larger leaves with blades 20–45 mm wide; leaf tip with an abrupt, short, stiff point; fruit gray-green to olive-green, the lateral keels without teeth…………………………………………………. *P. illinoensis*

33. Leaf blades sessile, not clasping the stem..34
33. Leaf blades with base conspicuously clasping the stem..35
34. Leaves with 3–9 veins, with 1–2 rows of lacunae on either side of midvein; fruits 1.9–2.3 mm with beak 0.3–0.5 mm long...*P. gramineus*
34. Leaves with 7–19 veins, with 2–5 rows of lacunae on either side of midvein; fruits 2.5–3.6 mm with beak 0.5–0.8 mm long...*P. illinoensis*
35. Stipules conspicuous and persistent; leaf blades often >10 cm long, toothless, the tips boat-shaped (hood-like); rhizomes rusty-red spotted; fruit 4–5.7 mm wide with a prominent keel; flowering stems usually >18 cm long...*P. praelongus*
35. Stipules inconspicuous or soon disintegrating; leaf blades usually <10 cm long, with minute, 1-celled teeth on margins, the tips flat; rhizomes unspotted; fruit 1.6–4.2 mm wide with an inconspicuous keel or without a keel; flowering stems usually <10 cm long.........36
36. Leaf blades egg-shaped to near circular, rarely to 5 cm long, delicately 7–15 veined; stipules delicate, disintegrating, and absent on lower part of stem; fruit 1.6–3 mm wide.... ....*P. perfoliatus*
36. Leaf blades narrowly egg-shaped to narrowly lance-shaped, mostly 3–10 cm long, coarsely 13–21 veined; stipules disintegrating into coarse, white fibers; fruit 2.2–4.2 mm wide..*P. richardsonii*

P. perfoliatus

P. gemmiparus

BIG-LEAVED PONDWEED *Potamogeton amplifolius* Tuck.

Big-leaved pondweed is a robust, submersed, rooted, perennial, with round, often rusty-spotted, few-branched stems (6–110 cm long) and creeping rhizomes. Leaves of two types: all submersed or both submersed and floating. Submersed leaves are alternate, with narrow, black-spotted stalks (0.5–4.5 cm long), lax, distinctly arching, often folded, broadly elliptic (5–12.5 cm long, 2.5–7.5 cm wide), undulating margins, green, often turning reddish-brown, 19–49 veins, lacunae absent, and pointed at tip. Stipules are conspicuous (1.5–11.7 cm long), light brown, sheathing, and completely free from leaf base. Floating leaves are stalked (stalks 2.3–22.6 cm long), blades light green, elliptic to narrowly egg-shaped (4.3–9.2 cm long, 2.5–3.8 cm wide), 27–49 veins, and rounded to lobed at base. Flower clusters are dense, cylindrical (3.4–6.5 cm long) and emersed above water (flowering stem 4.5–22 cm long, erect, axillary or terminal). Fruit reddish brown, 3.9–5.2 mm wide, with a keel on back and sides, and an erect beak (0.5–0.8 mm). Turions are not produced.

HABITAT & DISTRIBUTION

Potamogeton amplifolius is a native species of shallow to deep, lakes, ponds, and rivers. It is widespread in CT and RI. An easily recognized pondweed with its large, often reddish, strongly arching, many-veined, submerged leaves, and ability to colonize deeper waters.

Notes: Similar to Spotted pondweed (*P. pulcher*) but with rusty-spotted stems and larger, arching, many-veined, submersed blades. When floating leaves are absent, plants can be confused with Illinois pondweed (*P. illinoensis*). Known to naturally hybridize with *P. gramineus* and *P. illinoensis* in CT.

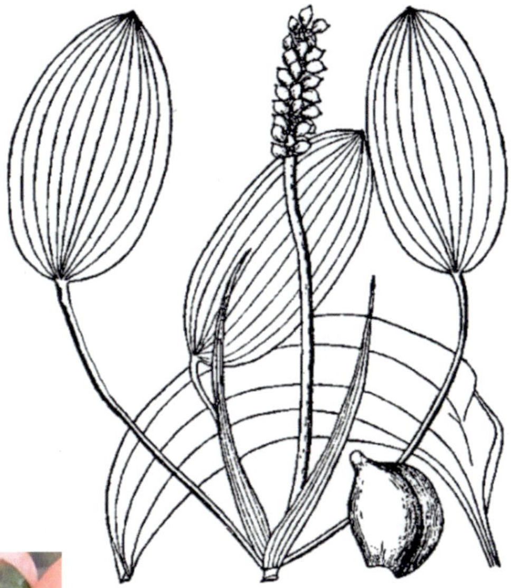

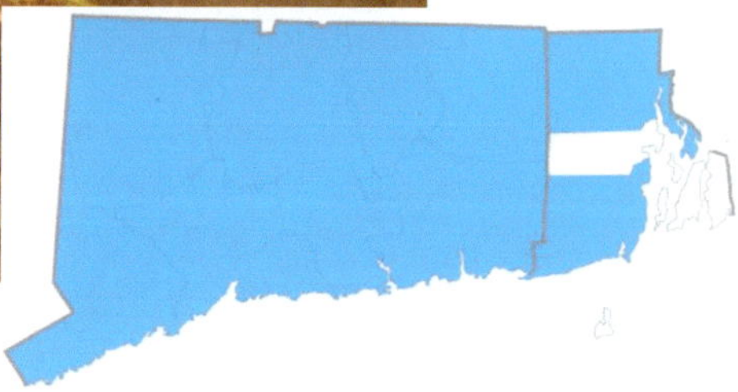

Berchtold's pondweed *Potamogeton berchtoldii* Fieber

Berchtold's pondweed is a delicate, submersed, rooted, perennial, with slender stems (18–150 cm long). Leaves are all submersed, alternate, flaccid, linear (0.9–5.4 cm long, 0.2–2.5 mm wide), pale green to olive-green, 1–5 veins, lacunae present in 1–5 rows on each side of vein, and pointed at tip. Paired, circular, raised glands (–0.5 mm wide), green (same as stem), are inconspicuous at a few nodes. Stipules are inconspicuous, 3.1–9.2 mm long, brown to green or white, delicate, not shredding at tip, wrapped around the stem with overlapping (unfused) margins, and completely free from leaf base. Flower clusters are mostly knob-like (1.5–10 mm long), with 1–3 crowded whorls of flowers/fruits, emersed above water or submersed, at end of cylindric stems (0.5–3.0 cm long), more than 3 per plant. Fruit green to brown, 1.5–2.2 mm wide, not keeled, sides rounded, and with an erect beak (0.1–0.6 mm). Turions are common, slender (0.9–3.2 cm long, 0.3–1.8 mm wide), soft, and leaves two-ranked, with inner leaves modified into a hardened structure.

Habitat & Distribution

Potamogeton berchtoldii is a native species of shallow lakes, pools, and streams. It is widespread and common in CT and RI.

Notes: This is our most common narrow-leaved pondweed. It is similar to Small pondweed (*P. pusillus*), but with more (1–5 rows) lacunae per side of leaf midrib, >3 flowering stems per plant, and uninterrupted flowering spikes.

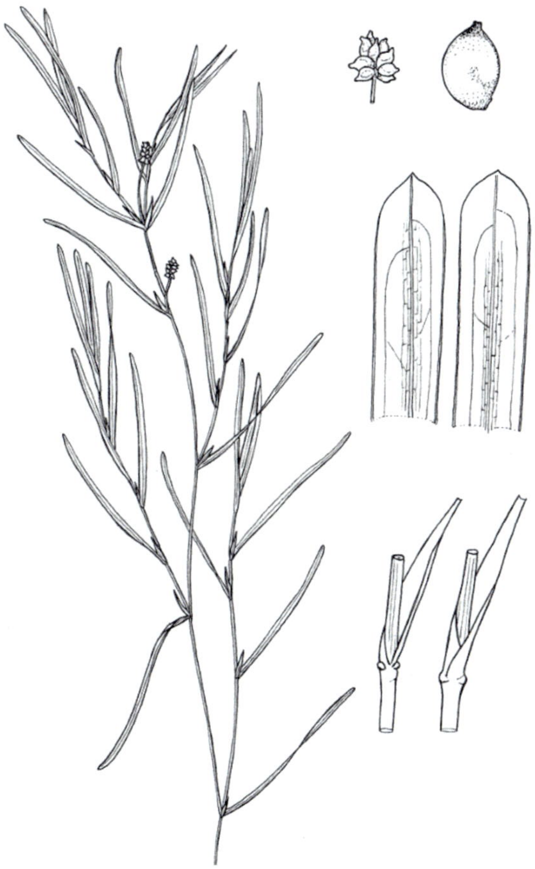

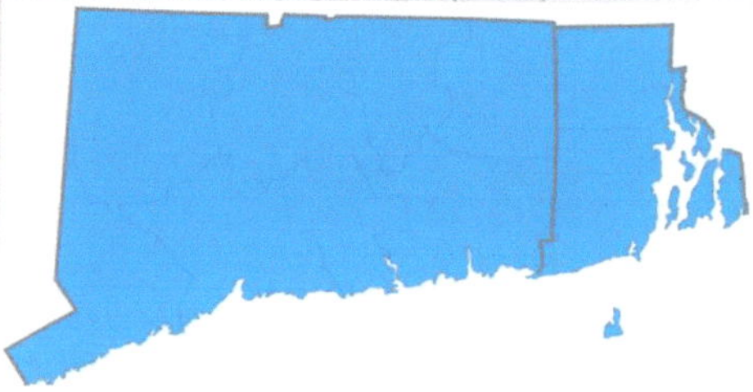

SNAIL-SEED PONDWEED *Potamogeton bicupulatus* Fern.

Snail-seed pondweed is a submersed, rooted, perennial, with compressed, highly branched, slender stems (10–25 cm long) and creeping rhizomes. Leaves of two types: all submersed or both submersed and floating. Submersed leaves alternate, linear and thread-like (1.5–11 cm long, 0.1–0.4 mm wide), light green, 1 vein, lacunae absent, and pointed or long-tapering at tip. Stipules are inconspicuous, light green, fused to base of submersed leaves for <½ of stipule length, with remaining tip of stipule free (usually greater than fused portion). Floating leaves are stalked (stalks 3–35 mm long), blades elliptic to lance-shaped (6–23 mm long, 1–11 mm wide), 3–7 veins, with acute to long-tapering leaf tips. Flower clusters of two types: either globular (1.5–7 mm long) and submersed (flowering stem 1–10 mm long, somewhat recurved, axillary) or short-cylindrical (3–14 mm long) and emersed above water (flowering stem 4–22 mm long, erect or recurved, axillary or terminal). Fruit greenish brown, 1–1.5 mm wide (including keel), with well-developed bumpy keel and usually lacking a beak. Embryo distinctly coiled with more than a full spiral. Turions are not produced.

HABITAT & DISTRIBUTION

Potamogeton bicupulatus is a native, northeastern species of shallow, usually acidic, lakes and streams. It is widespread in CT and RI. Known to naturally hybridize with Ribbonleaf pondweed (*P. epihydrus*) in CT.

Notes: This is one of only two species in our flora that produces two types of flowering spikes (emersed and submersed). The other species is Northern snail-seed pondweed (*P. spirillus*) which is similar to *P. bicupulatus* but has ribbon-like submersed leaves, fused portion of stipules about half stipule length, and rounded floating blades.

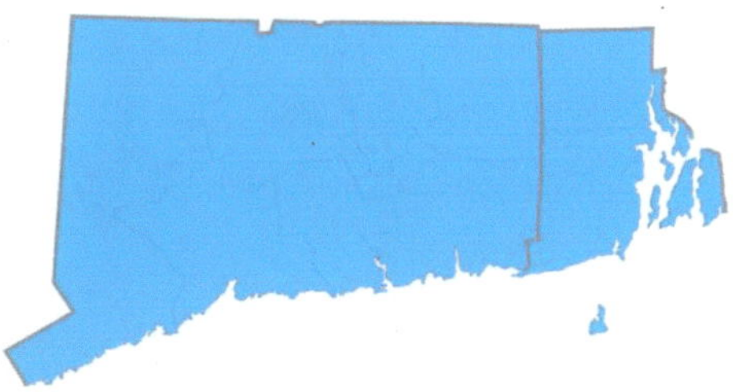

ALGA PONDWEED *Potamogeton confervoides* Rchb.

Alga pondweed is a very delicate submersed, rooted, perennial, with round, much-branched stems (10–80 cm long) and obvious, elongated, creeping rhizomes. Leaves alternate, submersed, thread-like (2–6.5 cm long, 0.1–1 mm wide), flaccid, pale green, 1 vein, and long-tapering and bristly at tip. Stipules are inconspicuous, pale green, and completely free from leaf base. Flower clusters short (5–12 mm long), emersed above water, at end of unusually long stems (5–25 cm long). Fruit light green, 2–3 mm long, 1.7–2.8 mm wide, with keels and erect beak (0.5 mm). Turions (0.7–2 cm long) present in leaf axils of old leaves.

HABITAT & DISTRIBUTION

Potamogeton confervoides is a northern, native species of shallow, acidic, lakes and ponds. It is rare and state-listed as ***endangered*** in CT (***historic*** in RI).

Notes: This plant is found in our most acidic waters (≤ pH 5) including bog ponds. Its leaves are so delicate and hair-like that when the plant is removed from the water, they collapse onto each other. Easy to overlook if not flowering.

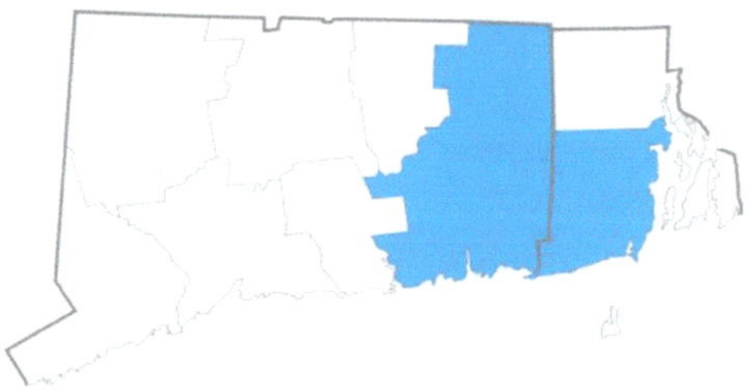

Curly Pondweed *Potamogeton crispus* L.

Curly pondweed is a submersed, rooted, perennial, with flattened, few-branched, slender stems. Leaves are all submersed, alternate, bright green, slightly stiff, linear-oblong (12–90 cm long, 4–10 cm wide), round at tip, with 3–5 veins and 2–5 rows of lacunae on each side of midvein. Leaf margins are wavy and conspicuously sharp toothed especially towards tip. Stipules are brown, inconspicuous, and completely free from leaf base. Flowers are borne above water in cylindrical spikes (10–15 mm long) with 6–20 flowers. Fruits (6 mm long, 2.5 mm wide) are brown to red brown, with a prominent (2–3 mm long) curved beak and wavy keel. Turions (1.5–3 cm long, 2 cm wide), consisting of hard, rolled leaves, are commonly formed in leaf axils or at stem tips.

Habitat & Distribution

Potamogeton crispus is a non-native species of alkaline, often polluted, lakes and slow, freshwater to brackish rivers. It is **invasive** and banned in CT and RI.

Notes: Introduced from Europe, this species can be a problematic weed. It is an unusually early blooming species and produces turions in early summer. Plants typically go dormant and decay in late summer as turions germinate into small plants.

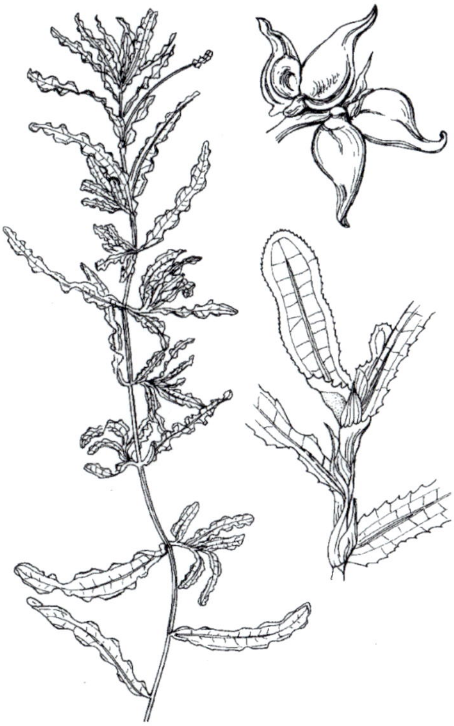

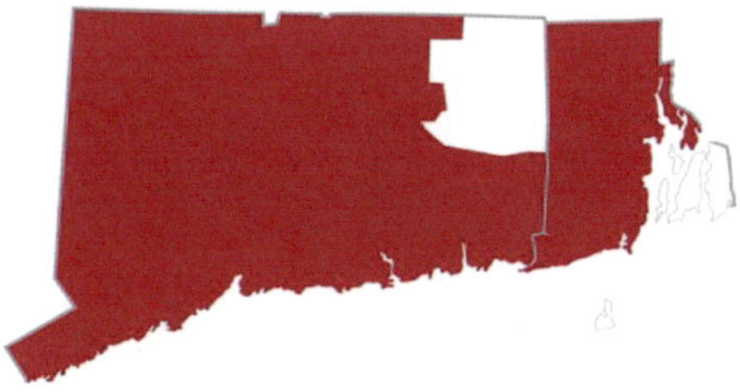

RIBBONLEAF PONDWEED *Potamogeton epihydrus* Raf.

Ribbonleaf pondweed is a submersed, rooted, perennial, with slightly flattened, slender stems (10–90 cm long) and creeping rhizomes. Leaves of two types: all submersed or both submersed and floating. Submersed leaves alternate, flat, linear (5–22 cm long, 2–10 mm wide), red-brown to light green, thin and transparent, 3–13 veined, with prominent lacunae bands 1–2 mm wide on each side of mid-vein, and rounded at tip. Stipules are inconspicuous, reddish brown to light green, sheathing, completely free from leaf base. Floating leaves are stalked (stalks 2–13 cm long), blades light green, elliptic (2.5–9 cm long, 4–20 mm wide), 11–41 veins, with rounded leaf tips. Flower clusters are dense, cylindrical (0.8–4 cm long) and emersed above water (flowering stem 1.5–5 cm long, erect, and axillary). Fruit greenish brown, 2.5–4.5 mm wide, with acute keels on back and sides and an erect beak (0.5 mm). Turions are not produced.

HABITAT & DISTRIBUTION

Potamogeton epihydrus is a native species of shallow, acidic or alkaline, lakes, ponds, and slow rivers. It is widespread and common in CT and RI.

Notes: Probably our most common pondweed. It is known to hybridize with Snailseed pondweed (*P. bicupulatus*) in CT.

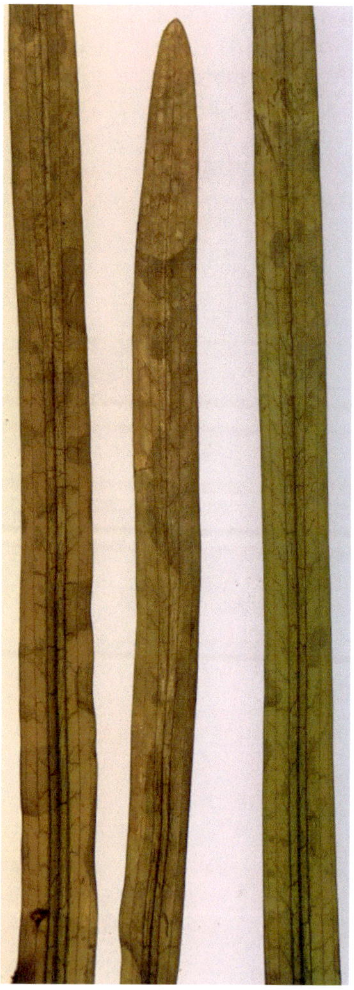

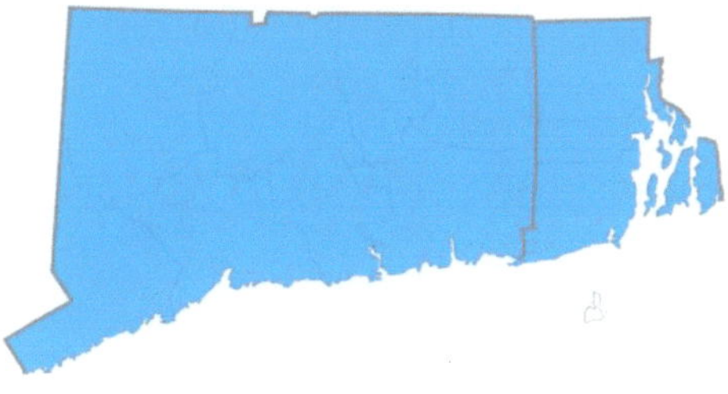

Leafy pondweed — *Potamogeton foliosus* Raf.

Leafy pondweed is a delicate, submersed, rooted, perennial, with much-branched, slightly compressed stems (4–75 cm long). Leaves are all submersed, alternate, flaccid, linear (1.3–8.2 cm long, 0.3–2.3 mm wide), pale green to olive-green, 3–5 veins, lacunae rarely present in 0–2 rows on each side of midvein, and pointed at tip (rarely with bristle). Circular glands (to 0.3 mm wide), black to gold, are rarely present at nodes. Stipules are inconspicuous, 2–22 mm long, brown to green, delicate, wrapped around the stem with overlapping margins, and completely free from leaf base. Flower clusters are knob-like to cylindric (1.5–7 mm long), emersed above water, at end of axillary, short (3–11 mm long) stems. Fruit olive-green to brown, 1.4–2.3 mm wide, with a wing-like, undulating keel on back, and erect beak (0.2–0.6 mm). Turions are uncommon, slender (0.9–2.5 cm long, 0.6–2 mm wide), soft, with inner leaves modified into a hardened structure.

Habitat & Distribution

Potamogeton foliosus is a native species of shallow lakes and streams. It is widespread in Ct and Ri.

Notes: This is the only narrow-leaved pondweed with an undulating, wing-like keel along back of the fruit.

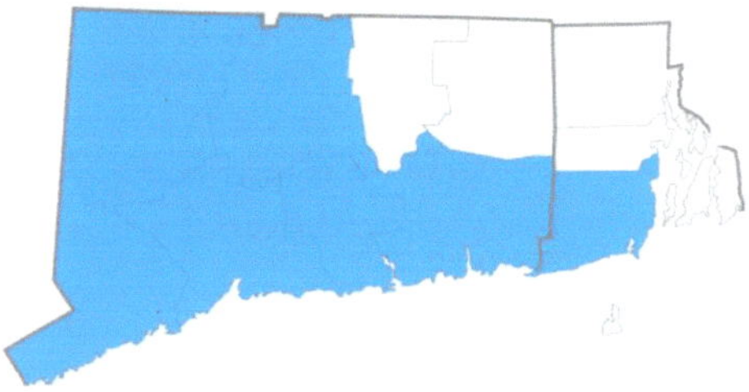

FLAT-STALKED PONDWEED *Potamogeton friesii* Rupr.

Flat-stalked, or Fries', pondweed is a delicate submersed, rooted, perennial, with compressed, slender stems (10–135 cm long). Leaves are all submersed, alternate, linear (2.3–6.5 cm long, 1.2–3.2 mm wide), light green, 5–7 veins, lacunae absent or present in 1 row on each side of midvein, and obtuse or rounded at tip, sometimes with a short projected tip. Circular raised glands (to 0.7 mm wide), green, greenish brown, or gold, are usually conspicuous at some nodes. Stipules are inconspicuous, 5–21 mm long, fibrous, white, coarse, shredding at tip, sheathing the stem, and completely free from leaf base. Flower clusters are cylindrical (7–16 mm long), dense, emersed above water, at end of flattened stems (1.2–4 cm long). Fruit olive green to brown, 1.8–2.5 mm long, lacking keels, and with an erect beak (0.3–0.7 mm). Turions are common, slender (15–50 mm long, 1.5–4 mm wide), soft, and leaves four-ranked, with inner leaves modified into a fan-shaped structure oriented at 90° angle to outer leaves, and outer leaves corrugated at base.

HABITAT & DISTRIBUTION

Potamogeton friesii is a native species of shallow, alkaline, lakes and ponds. It is very rare in CT where it is state-listed as ***endangered***.
Note: This is the only pondweed with turions where the outer leaves have corrugated bases and the inner leaves turned at right angles to the outer leaves. It is most similar to Straight-leaved pondweed (*P. strictifolius*).

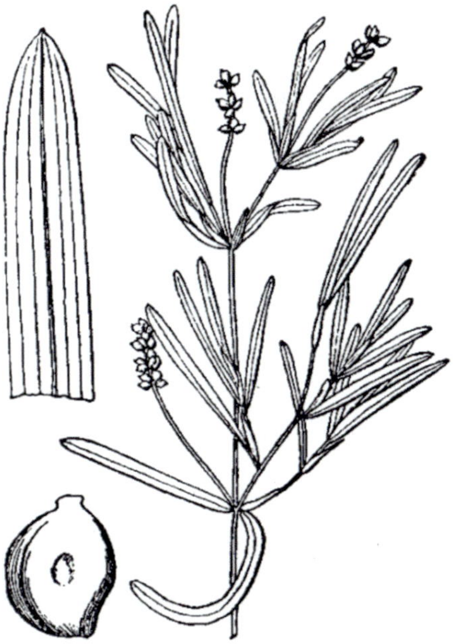

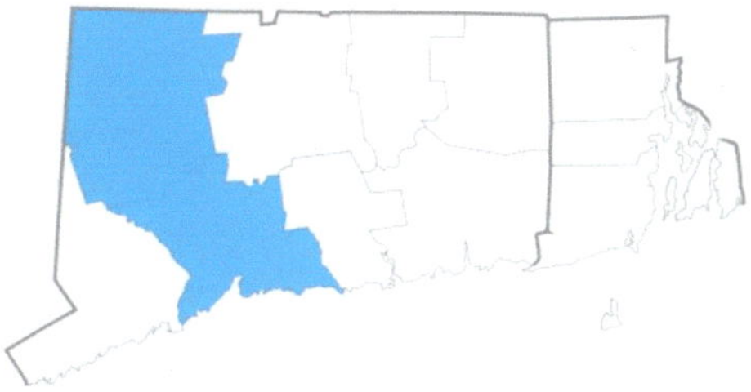

BUDDING PONDWEED *Potamogeton gemmiparus* (Robb.) JW Robbins

Budding pondweed is a very delicate submersed, rooted, perennial, with round, slender stems (18–150 cm long). Leaves are all submersed, alternate, linear and thread-like (1–6 cm long, 0.2–0.7 mm wide), pale- to olive-green, 1 vein, lacunae absent or present in 0–2 rows on each side of vein, and narrowly tapering to tip. Circular raised glands (to 0.5 mm wide), green, gold, brown, or rarely white, are present on at least some nodes. Stipules are inconspicuous, 5–20 mm long, brown to green or white, sheathing the stem, and completely free from leaf base. Flower clusters are short-cylindrical (1.5–10 mm long), dense and uninterrupted, submersed or emersed above water, at end of stems (1–3.5 cm long). Fruit green to brown, 1.5–2 mm long, lacking keels, and with an erect beak (0.1–0.6 mm). Turions are common as slender (10–30 mm long, 0.3–1.8 mm wide), soft, two-ranked outer leaves and rolled, hardened inner leaves.

HABITAT & DISTRIBUTION

Potamogeton gemmiparus is a native northern species of shallow, acidic waters of lakes, ponds, and slow rivers. It is rare and state-listed as ***threatened*** in CT and ***historical*** in RI.

Notes: This narrow-leaved pondweed is geographically restricted to New England and southern Quebec. It is difficult to distinguish from the more widespread Small pondweed (*P. pusillus*) or Berchtold's pondweed (*P. berchtoldii*).

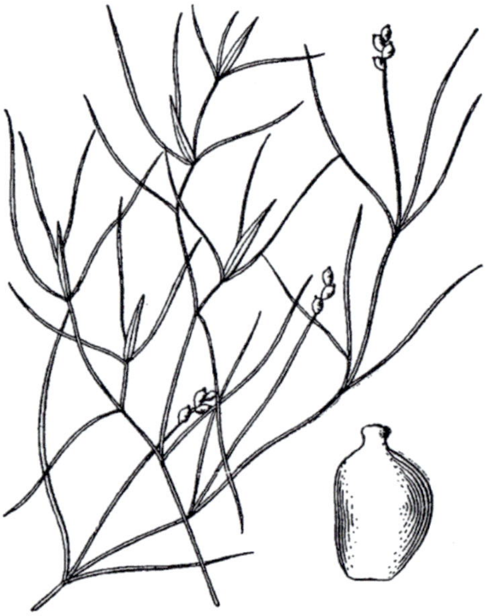

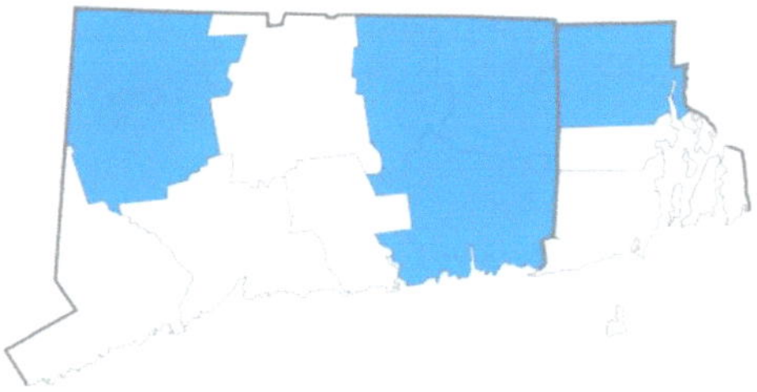

GRASSY PONDWEED *Potamogeton gramineus* L.

Grassy pondweed is a submersed, rooted, perennial, with round to flattened, slender, branching stems (to 150 cm long) and creeping rhizomes. Leaves of two types: all submersed or both submersed and floating. Submersed leaves are alternate, lax, elliptic (3.1–9.1 cm long, 3–27 mm wide), without stalks, light to brownish green, 3–9 veins, with 1–2 rows of lacunae on either side of midvein, and pointed at tip. Stipules are inconspicuous (1.3–1.6 cm long), pale green to brown, sheathing, and completely free from leaf base. Floating leaves are stalked (stalks 3–5.5 cm long, continuous in color to blade junction), blades yellow green to dark green, elliptic to egg-shaped (3.5–4 cm long, 16–20 cm wide), 11–13 veins, and rounded at base. Flower clusters are dense, cylindrical (1.5–3.5 cm long) and emersed above water (flowering stem 3.2–7.7 cm long, erect, axillary and terminal). Fruit greenish brown, 1.9–2.3 mm wide, with a keel on back and on sides, and an erect beak (0.3–0.5 mm). Turions are not produced.

HABITAT & DISTRIBUTION

Potamogeton gramineus is a native northern species of shallow lakes, ponds, and rivers. It is widespread in CT and RI. Leaf shape, size, and arrangement can be highly variable in relation to water depth.

Notes: Similar to the northern, cold-water, Alpine pondweed (*P. alpinus* Balbis)—state-listed as ***endangered*** in CT—yet suspected to no longer occur there. Alpine pondweed differs by having red-tinged, longer (to 18 cm), submersed leaves, unbranched stems, and larger, plump fruits. Grassy pondweed is known to naturally hybridize with *P. amplifolius*, *P. illinoensis*, and *P. perfoliatus* in CT.

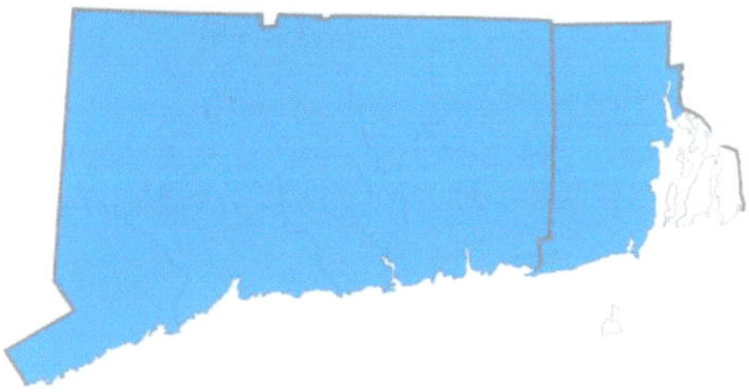

HILL'S PONDWEED *Potamogeton hillii* Morong

Hill's pondweed is a delicate, submersed, rooted, perennial, with very slender, much-branched, slightly compressed stems (30–60 cm long). Leaves are all submersed, alternate, linear (2–6 cm long, 0.6–2.5 mm wide), pale green to olive-green, 3(–5) veins, lacunae present in 1–2 rows on each side of midvein, and sharp bristle-tipped (or with an abrupt, short, projected tip). Glands usually absent at nodes, brown to green (0.1–0.3 mm wide) if present. Stipules are inconspicuous, 7–16 mm long, white to light brown, wrapped around the stem with overlapping margins, and completely free from leaf base. Flower clusters are nearly spherical (4–7 mm long), emersed above water, at end of axillary and terminal, short (6–14 mm long), recurved stems. Fruit brownish, 2.3–4 mm wide, 3-keeled, and erect beak (0.3–0.7 mm). Turions are uncommon, slender (2.8–3 cm long, 1.5–3 mm wide), soft, with leaves 2-ranked.

HABITAT & DISTRIBUTION

Potamogeton hillii is a native species of shallow, mostly alkaline, lakes and streams. It is rare in CT where it is state-listed as ***endangered***.

Notes: This is the only narrow-leaved pondweed possessing leaves with <5 veins, bristle-like tips, and no nodal glands. It is similar to Leafy pondweed (*P. foliosus*).

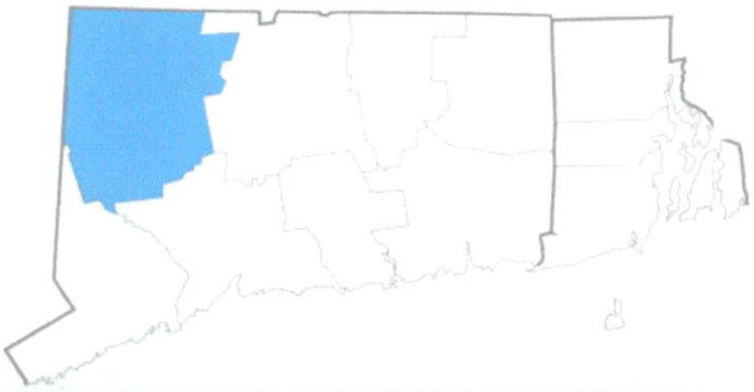

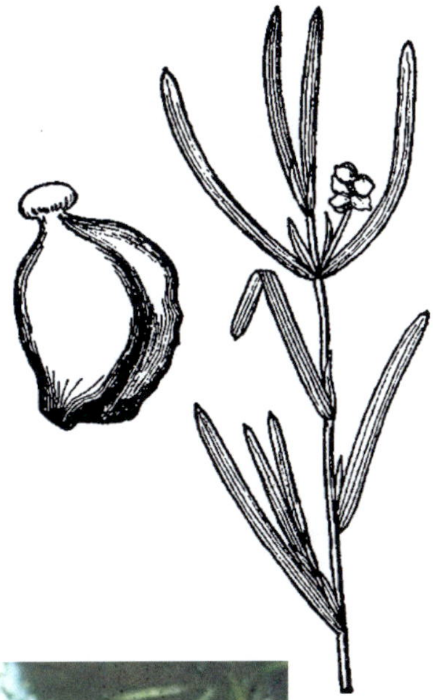

ILLINOIS PONDWEED *Potamogeton illinoensis* Morong

Illinois pondweed is a submersed, rooted, perennial, with round, few-branched, stems (28–120 cm long), and creeping rhizomes. Leaves of two types: all submersed or both submersed and floating. Submersed leaves are alternate, stalkless or with short stalks (0.5–4 cm long), lax, flat to undulating, lance-shaped to elliptic (5–20 cm long, 20–45 mm wide), red-brown to light green, 7–15 veins, with 2–5 rows of lacunae on each side of a prominent mid-vein, and abruptly tapered to a short (to 4 mm), stiff point at tip. Stipules are conspicuous (1–8 cm long), light brown to red-brown, sheathing, and completely free from leaf base. Floating leaves are stalked (stalks 2–9 cm long), blades light green, elliptic to oblong (4–19 cm long, 2.0–6.5 cm wide), 13–29 veins, and tapering at base. Flower clusters are dense, cylindrical (2.5–7 cm long) and emersed above water (flowering stem 4–30 cm long, erect, axillary or terminal). Fruit gray green to olive-green, 2.5–3.6 mm wide, with well-developed keels and an erect to slightly curved beak (0.5–0.8 mm). Lateral keels without teeth. Turions are not produced.

HABITAT & DISTRIBUTION

Potamogeton illinoensis is a stout, native species of shallow, mostly alkaline, lakes, ponds, rivers, and streams. It is widespread in CT.
Notes: Floating leaves are often not produced, especially in deeper water. It is similar to Long-leaf pondweed (*P. nodosus*), but with stalkless or shorter-stalked, wider, submersed leaves and green fruits. When submersed blades are stalkless, plants resemble Grassy pondweed (*P. gramineus*) but have more numerous veins and larger fruits. It is known to hybridize with both *P. gramineus*, *P. amplifolius*, and *P. richardsonii* in CT.

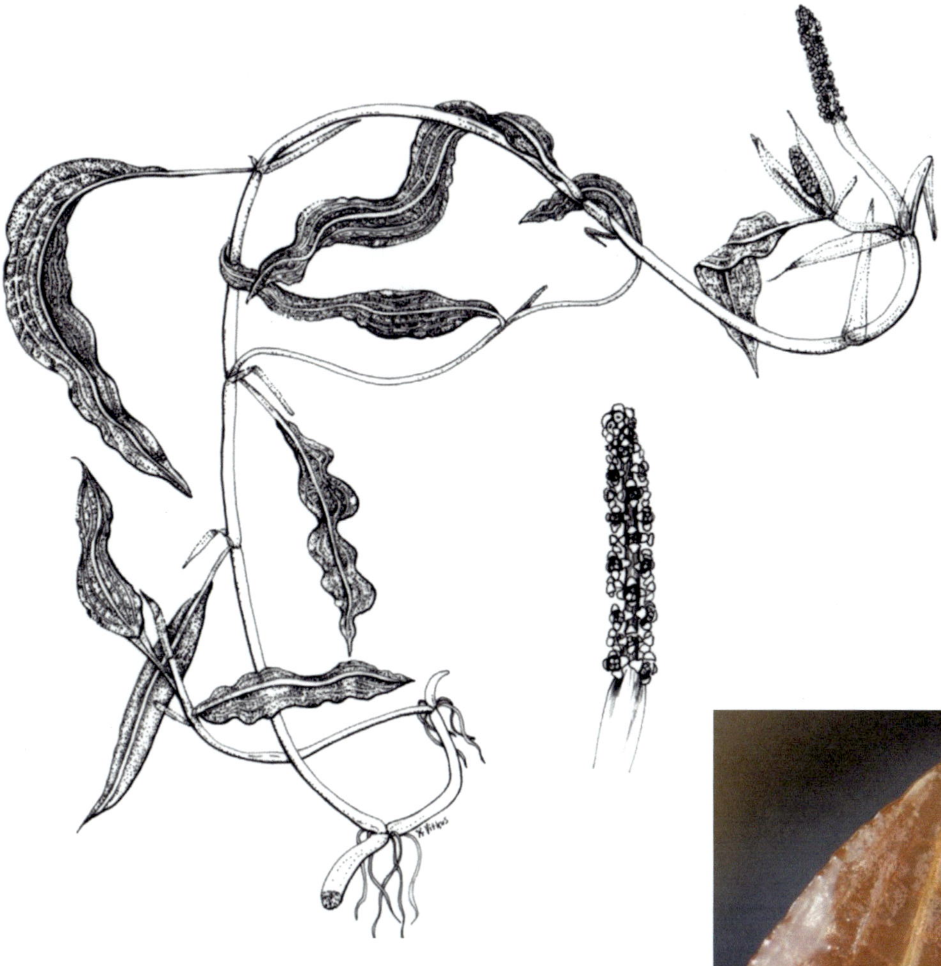

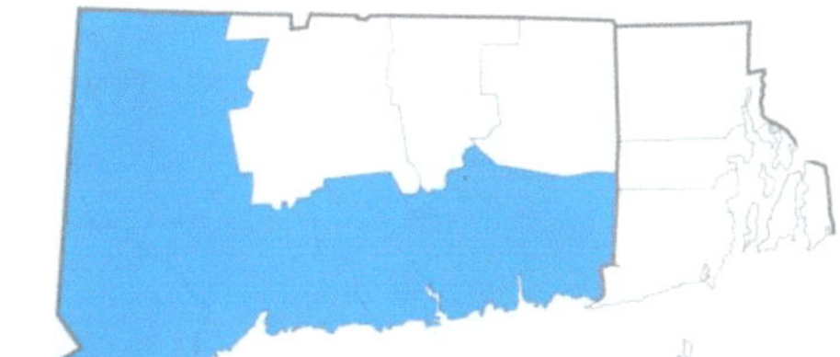

FLOATING PONDWEED *Potamogeton natans* L.

Floating pondweed is a submersed, rooted, perennial, with round, mostly unbranched, slender stems (30–90 cm long) and creeping rhizomes. Leaves both submersed and floating. Submersed leaves are alternate, stiff, linear (9–20 cm long, 0.7–2.5 mm wide), phyllodial (appearing as a bladeless leaf stalk), light to dark green, 3–5 obscure veins, lacking lacunae bands, and blunt at tip. Stipules are conspicuous (4–10 cm long), whitish, sheathing, and completely free from leaf base. Floating leaves are long-stalked (stalks 5–29 cm long, with a distinctly bent and lighter area at blade junction), blades light green, oval-elliptic to egg-shaped (5–10 cm long, 1.5–6 cm wide), 17–35 veins, usually rounded or heart-shaped at base. Flower clusters are dense, cylindrical (2.5–5 cm long) and emersed above water (flowering stem 4.5–9.5 cm long, erect, and terminal). Fruit greenish brown, 3.5–5 mm wide, wrinkled on sides, with an inconspicuous keel on back and an erect to curved beak (0.4–0.8 mm). Turions are not produced.

HABITAT & DISTRIBUTION

Potamogeton natans is a native species of acidic to alkaline lakes, ponds, rivers and streams, including tidally influenced waters. It is widespread and common in CT and RI.

Notes: This plant is easily identified by its phyllodial submersed leaves (when persisting), a band of paler green tissue on end of leaf stalk where it meets a floating blade, and usually lobed base of floating blades. In non-flowing waters, the apex of leafstalks on floating leaves are noticeably bent at a 90° angle where it meets the blade. The floating leaf blade shape and degree of lobing at base is influenced by water currents. It is similar to Oakes' pondweed (*P. oakesianus*).

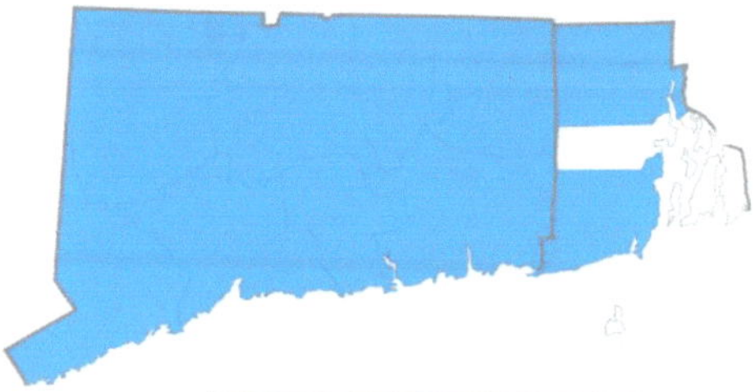

Long-leaf pondweed *Potamogeton nodosus* Poir.

Long-leaf pondweed is a submersed, rooted, perennial, with round, few-branched, stems (to 100 cm long), and whitish rhizomes. Leaves of two types: all submersed or both submersed and floating. Submersed leaves are alternate, with long stalks (5–15 cm long), lax, flat, narrowly lance-elliptic (9–20 cm long, 10–35 mm wide), light to dark green, 7–15 veins, with 2–5 conspicuous rows of lacunae on each side of a prominent mid-vein, and pointed at tip. Stipules are conspicuous (3–9 cm long), light brown, sheathing, and completely free from leaf base. Floating leaves are stalked (stalks 3.5–26 cm long), blades light green, elliptic (3–11 cm long, 1.5–4.5 cm wide), 11–21 veins, and tapering to rounded at base. Flower clusters are dense, cylindrical (2–7 cm long) and emersed above water (flowering stem stout, 3–15 cm long, erect, terminal). Fruit reddish brown, 2.7–4.3 mm wide, with well-developed keels and an erect beak. Lateral keels with blunt to sharp teeth. Turions are not produced.

Habitat & Distribution

Potamogeton nodosus is a native species of shallow to deep lakes, ponds, and streams. It is widespread in CT and RI, but occupies mostly low alkaline, fast-flowing waters in eastern areas and highly alkaline ponds and sluggish streams in western areas.

Notes: Plants are easily distinguished by the large, long-stalked submersed leaves and long-stalked, elliptic floating leaves, each tapering at both ends. It is similar to Illinois pondweed (*P. illinoensis*), but with longer-stalked, narrower, submersed leaves and reddish fruits. Submersed leaves often senesce before fruiting, so plants then can resemble Floating pondweed (*P. natans*).

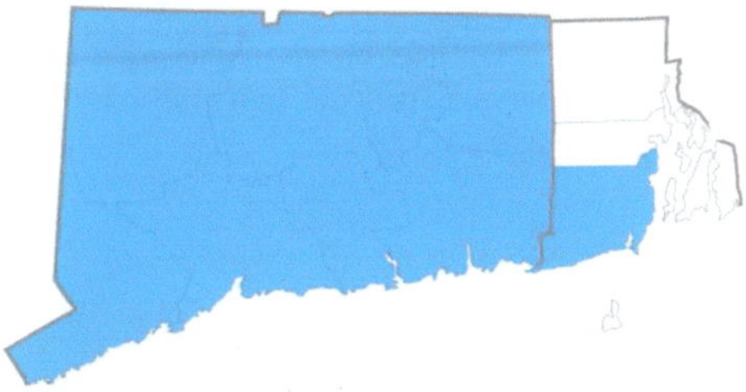

OAKES' PONDWEED *Potamogeton oakesianus* JW. Robbins

Oakes' pondweed is a submersed, rooted, perennial, with round, much-branched, red-spotted, slender stems (7–75 cm long) and creeping rhizomes. Leaves both submersed and floating. Submersed leaves are alternate, delicate and lax, linear (5–16 cm long, 0.25–1 mm wide), phyllodial (appearing as a bladeless leaf stalk), pale green, 3 veins, lacking lacunae bands, and pointed at tip. Stipules are conspicuous (2.5–4 cm long), whitish, sheathing, and completely free from leaf base. Floating leaves are stalked (stalks 3.2–7.5 cm long, continuous in color to blade junction), blades light to dark green, lance-to egg-shaped (2.5–6 cm long, 1–2 cm wide), 7–23 veins, and rounded or tapering at base. Flower clusters are dense, cylindrical (1–3.5 cm long) and emersed above water (flowering stem 2.5–8 cm long, erect, and terminal). Fruit greenish brown, 2–3.5 mm wide, not wrinkled on sides, with a prominent keel on back and keels on sides, and an erect beak (0.4–0.8 mm). Turions are not produced.

HABITAT & DISTRIBUTION

Potamogeton oakesianus is a native species of shallow, acidic, bogs, lakes, ponds and occasionally streams. It is widespread and common in CT and RI.

Notes: As our only other pondweed with submersed, phyllodial leaves, it is similar to Floating pondweed (*P. natans*), but these leaves in Oakes' pondweed are much narrower (delicate and soon deteriorating too) as are the floating blades, which are not lobed at base. This species is named after famous New England botanist William Oakes (d. 1848) of Ipswich, MA.

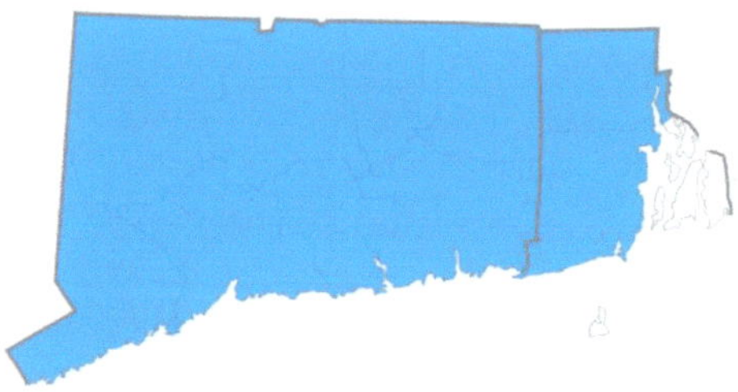

BLUNT-LEAVED PONDWEED *Potamogeton obtusifolius* Mert. & Koch

Blunt-leaved pondweed is a delicate, submersed, rooted, perennial, with slender, slightly compressed stems (35–90 cm long). Leaves are all submersed, alternate, flaccid, linear (3–8.2 cm long, 1–3.5 mm wide), green to reddish, 3 veins, lacunae present in 1–3 rows on each side of vein, and bluntly pointed or rounded at tip. Paired, circular, raised glands (0.3–1 mm wide), are yellowish green to gold, are common at nodes. Stipules are inconspicuous, 6–18 mm long, white, delicate, rarely shredding at tip, sheathing the stem, and completely free from leaf base. Flower clusters are cylindrical (8–13 mm long), emersed above water, at end of rounded stems (0.8–2 cm long). Fruit olive green to brown, 2.5–3.5 mm wide, sometimes keeled, and with an erect beak (0.8–1 mm). Turions are abundant, slender (3.5–7.8 cm long, 2.3–5.1 mm wide), soft, and leaves two-ranked, flattened, with inner and outer leaves in same plane.

HABITAT & DISTRIBUTION

Potamogeton obtusifolius is a native species of shallow, moderately calcareous waters of lakes and slow streams. It is infrequent in CT and RI.

Note: This narrow-leaved pondweed is very similar to the more common Berchtold's pondweed (*P. berchtoldii*), but with wider, blunt tipped leaves often flushed with maroon.

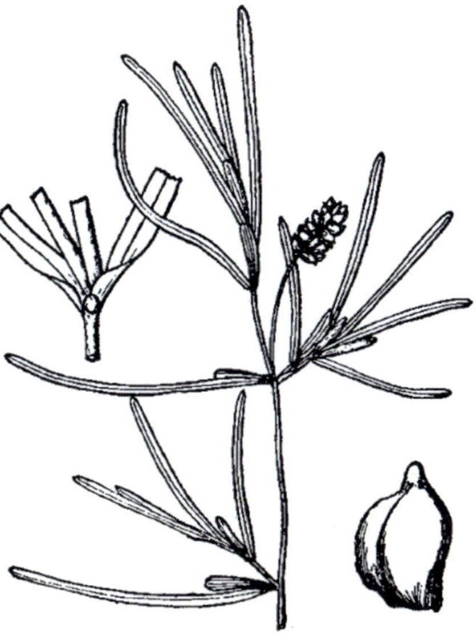

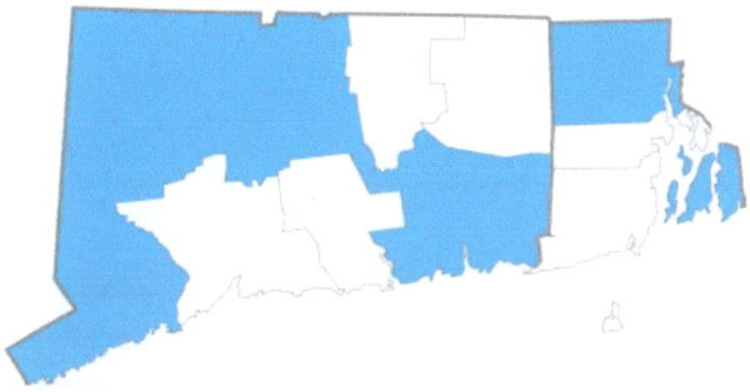

Clasping-leaved Pondweed *Potamogeton perfoliatus* L.

Clasping-leaved pondweed is a submersed, rooted, perennial, with round stems (to 2.5 m long) and unspotted rhizomes. Leaves submersed, alternate, lax, egg-shaped to near circular (1–7.6 cm long, 0.7–4 cm wide), with minute (1-celled) teeth on margins, blade base rounded and conspicuously clasping stem, olive green, 7–15 delicate veins, lacking lacunae, and with a flat, rounded tip. Stipules are inconspicuous (3.5–6.5 cm long), delicate, disintegrating into fibers, absent on lower part of stem, light brown to green, sheathing the stem, and completely free from leaf base. Flower clusters are cylindrical (0.4–4.8 cm long) and emersed above water, on cylindric stems (1–7.3 cm long). Fruit greenish brown, 1.6–3 mm wide, without keels, and an erect beak (0.4–0.6 mm). Turions are not produced.

Habitat & Distribution

Potamogeton perfoliatus is a native species of lakes, bays, and rivers, including brackish, tidal rivers. It is widespread in CT and RI.

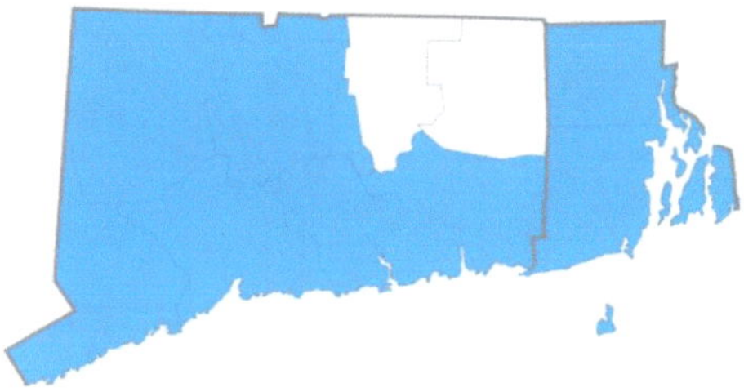

WHITE-STEMMED PONDWEED *Potamogeton praelongus* Wulfen

White-stemmed pondweed is a submersed, rooted, perennial, with round, zig-zagged stems (to 2 m long) and red spotted, white rhizomes. Leaves submersed, alternate, lax, linear to lance-shaped (6–28 cm long, 1–4.6 cm wide), blade base conspicuously clasping stem (without lobes), pale green, 11–33 veins, lacking lacunae, and with a hood-like, boat-shaped tip. Stipules are conspicuous (3–8 cm long), mostly white, fibrous, shredding at tip, sheathing the stem, and completely free from leaf base. Flower clusters are cylindrical (3.4–7.5 cm long) and emersed above water, on long (9.5–53 cm) stems. Fruit greenish brown, 4–5.7 mm wide, with a prominent, sharp keel on back, and an erect beak (0.6–0.1 mm). Turions are not produced.

HABITAT & DISTRIBUTION

Potamogeton praelongus is a native northern species of, shallow to deep, neutral to alkaline, waters of lakes and rivers. It is widespread in CT.

Notes: White-stemmed pondweed is distinctive with its large white stipules, submersed leaves clasping zig-zagged stems, and boat-shaped leaf tips (that split when pressed flat). It is known to hybridize with *P. richarsonii* in CT.

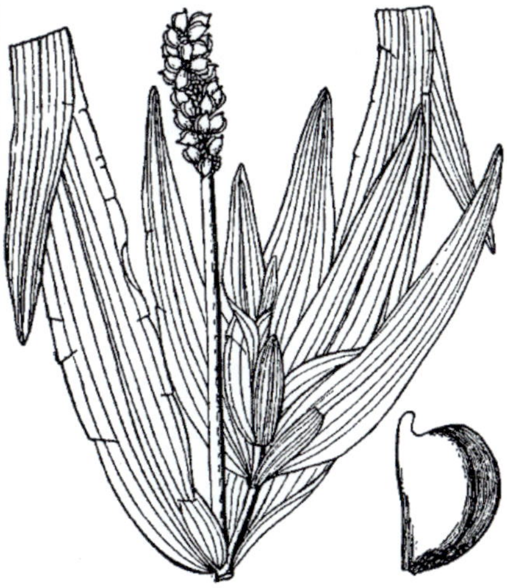

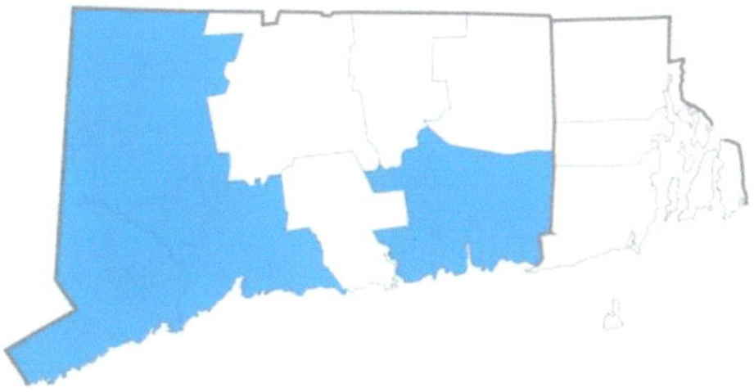

SPOTTED PONDWEED *Potamogeton pulcher* Tuck.

Spotted pondweed is a submersed, rooted, perennial, with round, conspicuously black-spotted, unbranched stems (8–95 cm long) and creeping rhizomes. Leaves both submersed and floating. Submersed leaves are alternate, with narrow, black-spotted stalks (0.5–4.5 cm long), lax, slightly arching, narrowly lance-shaped (3.5–13.8 cm long, 10–25 mm wide), undulating margins, dark green, 7–19 veins, with 2–5 rows of lacunae on either side of midvein, and pointed at tip. Stipules are inconspicuous (0.7–1.2 cm long), light to dark brown, sheathing, and completely free from leaf base. Floating leaves are stalked (stalks stout, 1–16.5 cm long, black-spotted), blades light to dark green, broadly elliptic (2.5–8.5 cm long, 1.1–4.4 cm wide), 15–19 veins, and rounded at base. Flower clusters are dense, cylindrical (1.7–3.6 cm long) and emersed above water (flowering stem 3.3–9.4 cm long, erect, axillary and terminal). Fruit dark greenish brown, 3.1–4.1 mm wide, with a keel on back and sides, and an erect beak (0.5 mm). Turions are not produced.

HABITAT & DISTRIBUTION

Potamogeton pulcher is a native, eastern species of shallow, acidic lakes and rivers. It is widespread in CT and RI. The conspicuously black-spotted stems and leaf stalks are distinctive.

Note: Similar to Big-leaved pondweed (*P. amplifolius*) but with black-spotted stems, less-broad submersed blades with fewer veins, and floating blades with fewer veins.

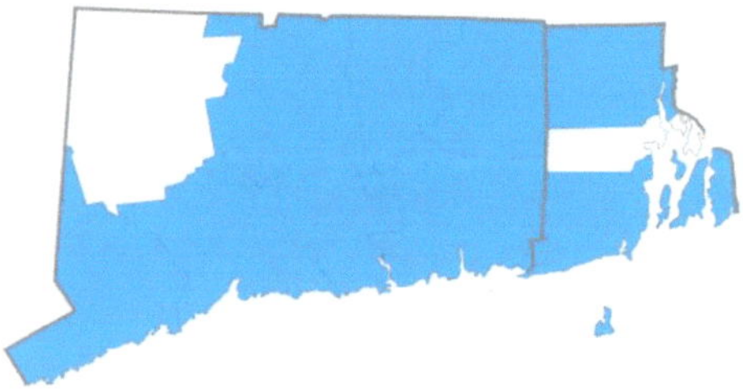

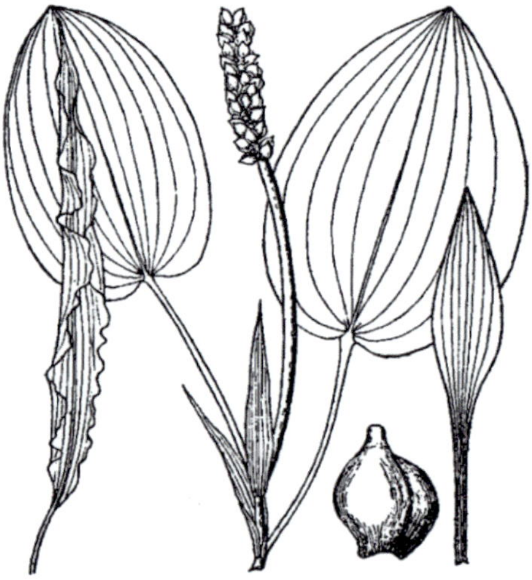

SMALL PONDWEED *Potamogeton pusillus* L.

Small pondweed is a delicate, submersed, rooted, perennial, with slender stems (18–150 cm long). Leaves are all submersed, alternate, flaccid, linear (1.4–6.5 cm long, 0.5–1.9 mm wide), pale green to olive-green, 1–5 veins, lacunae absent or present in 1–2 rows on each side of vein, and pointed at tip. Paired, circular, raised glands (to 0.5 mm wide), green (same as stem), are inconspicuous at a few nodes. Stipules are inconspicuous, 3.1–9.2 mm long, brown to green or white, delicate, not shredding at tip, with fused margins, forming a tube surrounding the stem, and completely free from leaf base. Flower clusters are cylindrical (1.5–10 mm long), usually with 2–4 interrupted whorls of flowers/fruits, emersed above water or submersed, at end of narrow, mostly terminal, stems (0.1–6.2 cm long), only 1–3 per plant. Fruit green to brown, 1.5–2.2 mm wide, not keeled, sides concave, and with an erect beak (0.1–0.6 mm). Turions are common, slender (0.9–3.2 cm long, 0.3–1.8 mm wide), soft, and leaves two-ranked, with inner leaves modified into a hardened structure.

HABITAT & DISTRIBUTION

Potamogeton pusillus is a native species of lakes, pools, and slow streams. It is uncommon in CT.

Notes: This narrow-leaved pondweed is distinctive with its narrow, 1–5 veined leaves, collar-like stipules, and interrupted fruit spikes (up to 3 per plant). It is similar to the more widespread Berchtold's pondweed (*P. berchtoldii*).

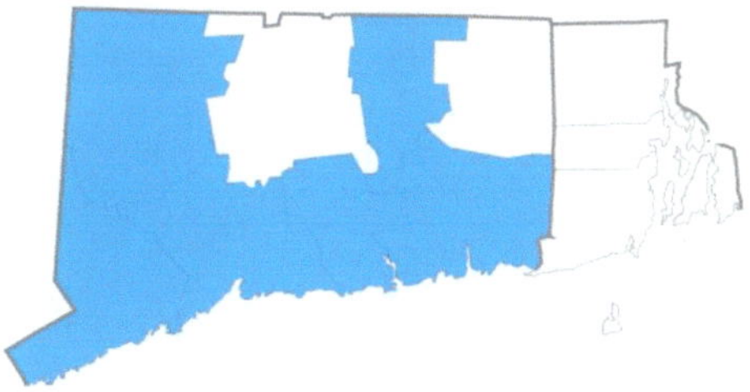

RICHARDSON'S PONDWEED *Potamogeton richardsonii* (Benn.) Rydb.

Richardson's pondweed is a submersed, rooted, perennial, with round stems (to 1 m long) and rhizomes. Leaves submersed, alternate, lax, narrowly egg-shaped to narrowly lance-shaped (1.6–13 cm long, 0.5–2.8 cm wide), with minute teeth on margins, blade base rounded and conspicuously clasping stem, olive green, 13–21 coarse veins, lacking lacunae, and with a flat, pointed tip. Stipules are conspicuous (1.2–1.7 cm long), firm, disintegrating into coarse, white fibers, sheathing the stem, and completely free from leaf base. Flower clusters are cylindrical (1.3–3.7 cm long) and emersed above water, on cylindric stems (1.5–14.8 cm long). Fruit greenish brown, 2.2–4.2 mm wide, usually without keels, and an erect beak (0.4–0.7 mm). Turions are not produced.

HABITAT & DISTRIBUTION

Potamogeton richardsonii is a native species of shallow, mostly calcareous, lakes and streams. It is uncommon in CT.

Notes: It is similar to the more widespread Clasping-leaved pondweed (*P. perfoliatus*). It is known to hybridize with *P. illinoensis, P. perfoliatus*, and *P. praelongus* in CT.

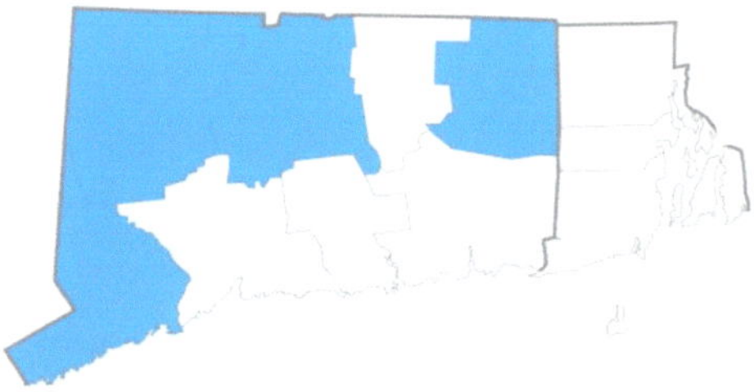

ROBBINS' PONDWEED *Potamogeton robbinsii* Oakes

Robbins' pondweed, or Fern pondweed, is a submersed, rooted, perennial, with round, branched, slender stems (to 1 m long) and creeping rhizomes. Leaves are all submersed, alternate, strongly two-ranked (attached on opposite sides of the stem so that whole plant is flat), closely spaced, dark green to reddish, stiff, lance-shaped (20–120 mm long, 3–8 mm wide), sharply pointed, and with pair of round lobes at base (where leaf meets stipule) that protrude past the stem. Leaves are often arching, with a conspicuous midvein, and finely toothed edges. A conspicuous, translucent, greenish brown to white, stipular sheath is fused to base of leaves for ¼ of stipule length, with remaining tip of stipule free and shredding. Flowers are borne just above water in slender spikes (7–20 mm long) on branching stems. Fruits (3–5 mm long, 2–3 mm wide), rarely produced, are brown, with an erect beak (0.7–0.9 mm long) and wavy keel. Turions are not produced.

HABITAT & DISTRIBUTION

Potamogeton robbinsii is a native species of lakes and slow rivers. It is common and widespread in CT and RI.

Notes: Plants are often in deep waters (5–7 m) where populations are dense (to 1000 plants per m^2). It is the only pondweed to have branched flowering stems and lobed leaf bases. It rarely flowers or sets fruit.

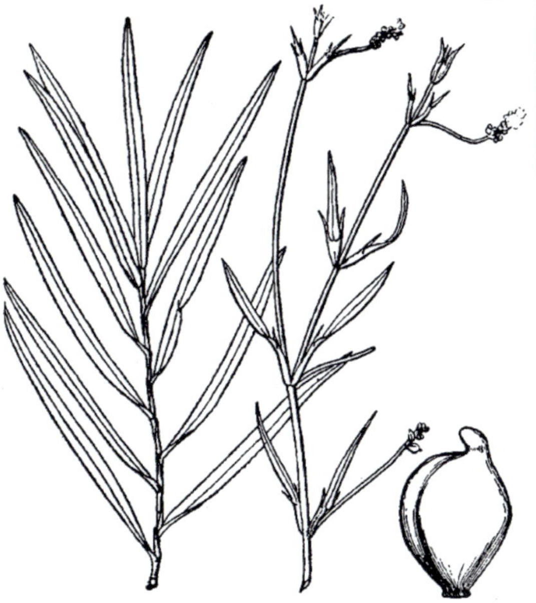

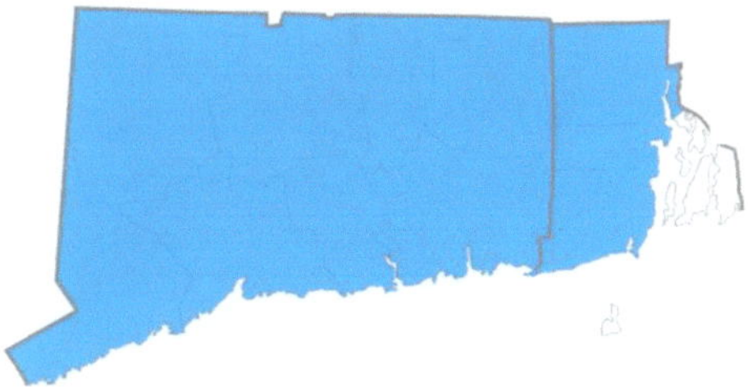

NORTHERN SNAIL-SEED PONDWEED *Potamogeton spirillus* Tuck.

Northern snail-seed pondweed is a submersed, rooted, perennial, with compressed, branched, slender stems (5–40 cm long) and creeping rhizomes. Leaves of two types: all submersed or both submersed and floating. Submersed leaves alternate, linear (0.7–80 cm long, 0.5–2 mm wide), red brown to light green, 1–3 veins, commonly with a broad lacunar band along each side of the midrib, and blunt-tipped. Stipules are inconspicuous, reddish brown to light green, fused to base of submersed leaves for ½ of stipule length, with remaining tip of stipule free (usually equal to or shorter than fused portion). Floating leaves are stalked (stalks 5–25 mm long), blades elliptic to lance-shaped (6–35 mm long, 2–13 mm wide), 5–15 veins, with rounded leaf tips. Flower clusters of two types: either globular (2–5 mm long) and submersed (flowering stem 0.5–3 mm long, recurved, axillary) or short-cylindrical (4–13 mm long) and emersed above water (flowering stem 4–27 mm long, erect or recurved, axillary or terminal). Fruit greenish brown, 1.5–2.5 mm wide (including keel), with blunt keel and lacking a beak. Embryo distinctly coiled in a full spiral. Turions are not produced.

HABITAT & DISTRIBUTION

Potamogeton spirillus is a native, northeastern species of lakes and rivers. It is widespread and common in CT and RI. **Notes:** This is one of only two species in our flora that produce two types of flowering spikes (emersed and submersed). The other species is Snail-seed pondweed (*P. bicupulatus*) which is similar to *P. spirillus* but has thread-like submersed leaves, fused portion of stipules shorter than free portion, and pointed floating blades.

STRAIGHT-LEAVED PONDWEED *Potamogeton strictifolius* Benn.

Straight-leaved pondweed is a delicate submersed, rooted, perennial, with slender, rounded, rigid stems (27–95 cm long). Leaves are all submersed, alternate, fairly stiff, linear (1.2–6.3 cm long, 0.6–2 mm wide), green to olive-green, 3–5 veins, lacunae absent, and sharp-pointed to nearly bristle-tipped. Paired, circular, raised glands (to 0.3 mm wide), are white, green, greenish brown, or gold, commonly at nodes. Stipules are inconspicuous, 6–16 mm long, white, fibrous, shredding at tip, with fused margins forming a cylinder or tube around stem, and completely free from leaf base. Flower clusters are cylindrical (6–13 mm long), emersed above water, at end of rounded stems (1–4.5 cm long). Fruit green-brown, 1.9–2.1 mm wide, not keeled, and with an erect beak (0.5–0.8 mm). Turions are common, slender (2.5–4.8 cm long, 1–2.2 mm wide), soft, and leaves two-ranked, flattened, with inner and outer leaves in same plane.

HABITAT & DISTRIBUTION

Potamogeton strictifolius is a native species of shallow, calcareous waters of lakes and slow streams. It is very rare in CT where it is state-listed as ***endangered***.

Note: The rigid leaves of this narrow-leaved pondweed hold their shape when removed from water.

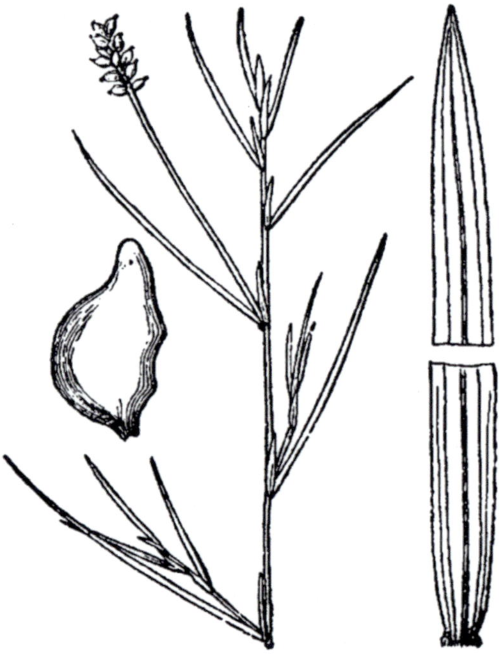

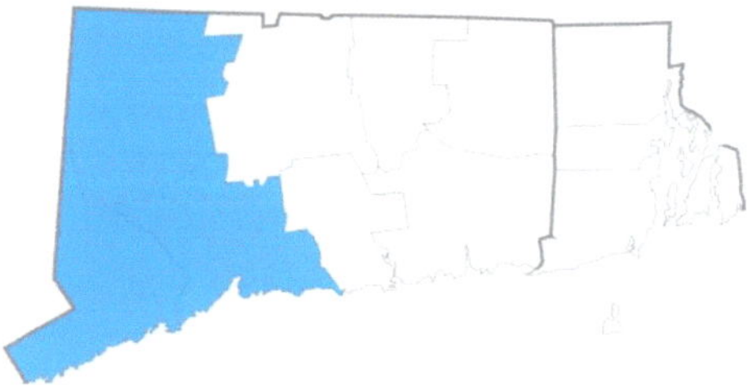

Vasey's pondweed — *Potamogeton vaseyi* J.W. Robbins

Vasey's pondweed is a very delicate submersed, rooted, perennial, with round, slender stems (10–25 cm long). Leaves of two types: all submersed or both submersed and floating. Submersed leaves alternate, linear and thread-like (2–8 cm long, 0.1–1 mm wide), light green, 1(–3) vein, lacunae absent, and pointed or long-tapering at tip. Stipules are inconspicuous, green to brown, and completely free from leaf base. Floating leaves are stalked (stalks 5–25 mm long), blades elliptic to oblong, small (6–15 mm long, 3–8 mm wide), 5–9 veins, with blunt leaf tips. Flower clusters short-cylindrical (3–8 mm long) at end of stems (5–30 mm long), emersed above water, and erect (or recurved in fruit). Fruit is greenish brown, 1.5–2.5 mm long, 1–1.5 mm wide, with a smooth keel and erect beak (0.3–0.5 mm), and embryo distinctly coiled with a full spiral. Turions are common in leaf axils as slender (5–20 mm long, 0.5–1.2 mm wide), soft, two-ranked outer leaves and rolled, hardened inner leaves.

Habitat & Distribution

Potamogeton vaseyi is a native species of shallow, acid to alkaline, lakes and rivers. It is rare and state-listed as ***threatened*** in CT.

Notes: Floating leaves are present only when plants are flowering and then plants deteriorate by mid-August. When floating leaves are absent, this plant is difficult to distinguish from Small pondweed (*P. pusillus*) or Budding pondweed (*P. gemmiparus*) with which it often co-occurs.

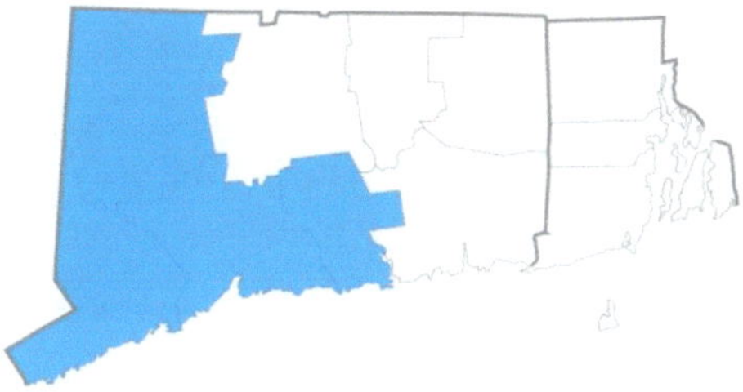

FLAT-STEM PONDWEED *Potamogeton zosteriformis* Fern.

Flat-stem pondweed is a submersed, rooted, perennial, with conspicuously flattened (appearing winged), stiff, stems (60–120 cm long, 0.6–3.2 mm wide). Leaves are all submersed, alternate, linear (10–20 cm long, 2–5 mm wide), light green, 15–35 veins, lacunae absent, and blunt to tapering at tip. Circular glands (to 0.3 mm wide), gold, are sometimes present at nodes. Stipules are conspicuous, 15–35 mm long, firm, fibrous, white, shredding at tip, sheathing the stem, and completely free from leaf base. Flower clusters are cylindrical (15–30 mm long), emersed above water, at end of stout, cylindric stems (2–5 cm long). Fruit olive green to green, 4–4.5 mm wide, sharp keeled on back, and with an erect beak (0.6–1 mm). Turions are common, flattened, large (4–7.5 cm long, 2–4.5 mm wide), firm, with leaves two-ranked, the inner and outer leaves oriented in the same plane, and the inner leaves undifferentiated

HABITAT & DISTRIBUTION

Potamogeton zosteriformis is a native species of shallow lakes, ponds, and streams. It is common in much of CT. It is a distinctive pondweed with its flat, winged stems and many-veined, parallel-sided, linear leaves to 5 mm wide. Leaves hold their shape when removed from water.

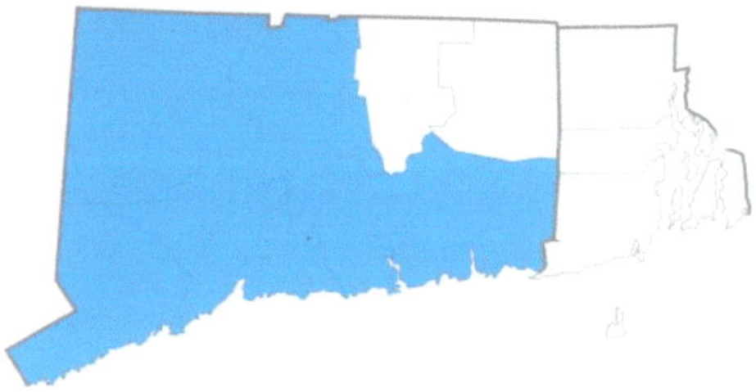

Proserpinaca – Haloragaceae

Mermaid-weed

The Mermaid-weeds are amphibious, rooted, perennials that can be mostly or entirely submersed. Stems, rooted from the lower nodes, grow up out of water from rhizomes. Leaves are alternate, and their form varies greatly by water depth: blades range from smaller and deeply pinnately divided (submersed) to larger and merely toothed (emergent). Flowers are small, purplish-greenish, and borne 1–3 in the emergent leaf axils. Fruit is a 3-seeded, 3-angled nutlet.

Two species occur in the region, both native.

KEY TO SPECIES

1. Emersed leaves (20–40 mm long) sharply-toothed, submersed leaves (5–30 mm long) deeply-divided into 7–14 pairs of segments; fruits 2.3–6 mm wide............................*P. palustris*
1. All leaves (10–30 mm long) deeply-divided, submersed leaves with 6–9 pairs of segments; fruits 2–2.8 mm wide.. *P. pectinata*

MARSH MERMAID-WEED *Proserpinaca palustris* L.

Marsh mermaid-weed is a submersed plant (10–40 cm long) with flowering portions emergent. Leaves are alternate (20–40 mm long), emersed ones sharply-toothed and submersed ones deeply-divided into 7–14 pairs of narrow segments 5–30 mm long. Flowers are borne in axils of emersed leaves, each with a toothed leaf-like bracts below. Fruits are 2.3–6 mm wide and sharply angled.

HABITAT & DISTRIBUTION

Proserpinaca palustris is native to shallow ponds and slow streams. It is widespread in CT and RI.

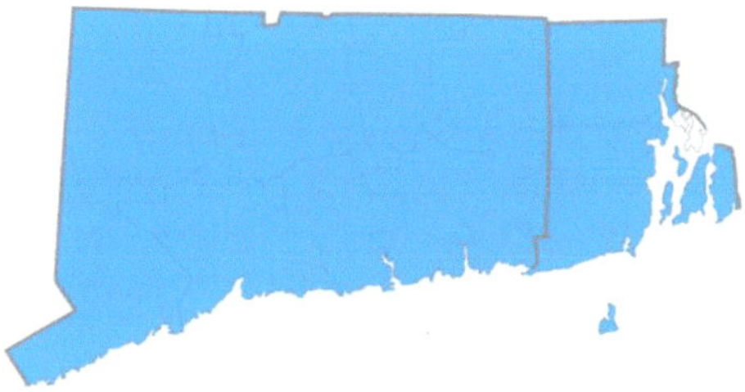

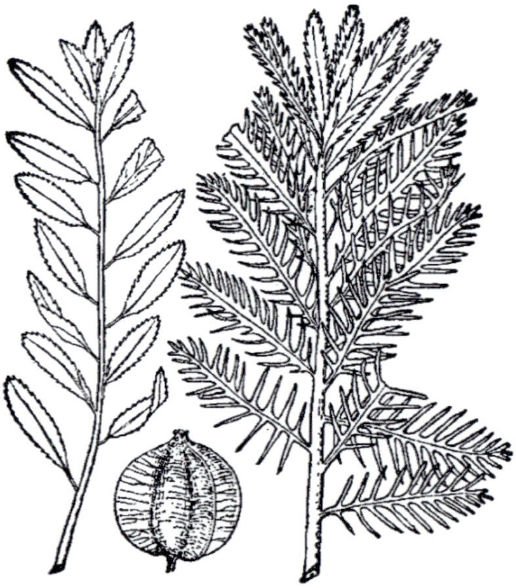

COMB-LEAVED MERMAID-WEED *Proserpinaca pectinata* Lam.

Comb-leaved mermaid-weed is usually a submersed plant with flowering portions emergent. Leaves are alternate (10–30 mm long) and deeply divided into 6–9 pairs of narrow segments 2–7.5 mm long. Flowers are borne in axils of emersed leaves, each with a deeply divided leaf-like bract. Fruits are 2–2.8 mm wide and bluntly angled.

HABITAT & DISTRIBUTION

Proserpinaca pectinata is native to shallow ponds, bogs, and ditches. This Coastal Plain species is rare and state-listed in RI as a species of **concern**.

Note: In RI, our two species of *Proserpinaca* are known to naturally hybridize to form hybrid plants (known as *P. ×intermedia* Mackenzie). These plants have submersed leaves with 7–12 pairs of short (2–3.5 mm) segments and fruits 2.3–3.6 wide.

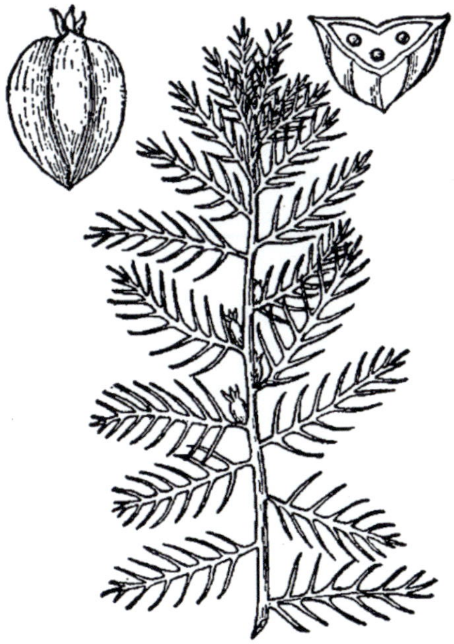

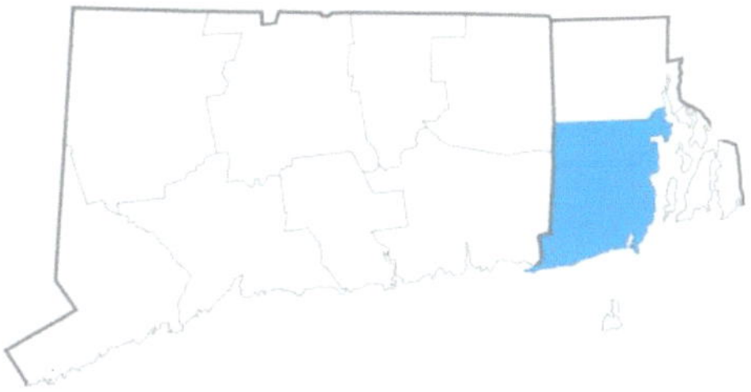

Ranunculus — Ranunculaceae
Crowfoot

The Crowfoots, or Buttercups, have just a few truly aquatic species with highly dissected, submersed leaves (or deeply lobed floating leaves). Leaves are obviously stalked and expanded at base sheathing the stem. Flowers are conspicuous and held above water, each with 5–6 colored petals, and numerous stamens and pistils. Fruits are clusters of small, seed-like achenes.

Three aquatic species occur in CT and RI, all native and highly variable in terms of leaf shape depending on conditions. In this region, there are several other species of Buttercups (e.g., *R. recurvatus* Poir., *R. repens* L., and *R. cymbalaria* Pursh) that occupy wetland shores. These emergent species (excluded here) have mostly erect stems and simple, broad leaves held above water.

KEY TO SPECIES

1. Leaves all arising from base of plant (basal) on long stalks, blades floating (or emersed), flat, and deeply lobed........ *R. sceleratus*
1. Leaves alternate along stems, mostly submersed and highly divided into very narrow segments........ 2
   2. Leaf segments thread-like, petals white (with yellow blotch at base); fruits transversely wrinkled........ *R. aquatilis*
   2. Leaf segments flat and ribbon-like; petals yellow; fruits smooth........ *R. flabellaris*

WHITE WATER CROWFOOT *Ranunculus aquatilis* L.

White Water crowfoot is a submersed perennial with elongate, branched stems rooted at lower nodes. Leaves are alternate, mostly submersed, round to semicircular in outline (2–3 cm long), highly divided into thread-like segments (less divided if floating), with expanded, sheathing stalks. Flowers are bisexual, emersed above water, borne singly on a long stalk from a leaf axil, with five white petals (4–7 mm long), and yellow centers. Fruit small (2 mm long) and transversely wrinkled.

HABITAT & DISTRIBUTION

Ranunculus aquatilis is a native species of ponds and slow streams. It is widespread in CT, but rare and state-listed as ***endangered*** in RI. Our plants are var. *diffusus* Withering.

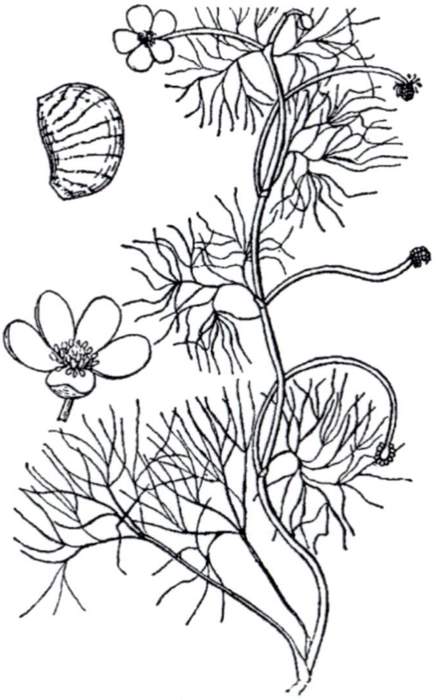

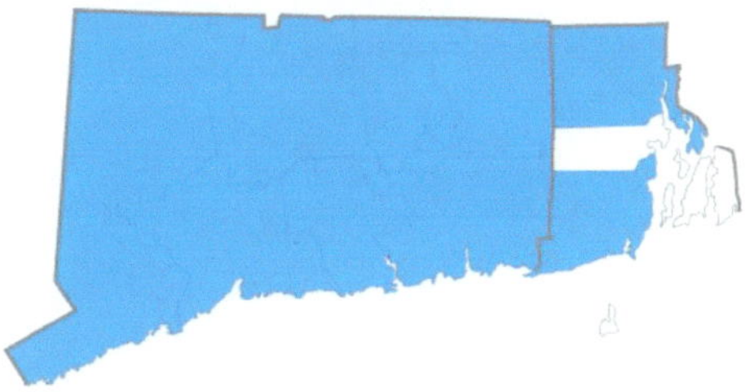

Yellow water crowfoot *Ranunculus flabellaris* Raf.

Yellow Water crowfoot is a submersed perennial with floating hollow stems rooted at lower nodes. Leaves are alternate, mostly submersed, kidney-shape to semicircular in outline (1.5–7 cm long), highly divided into flat, ribbon-like segments (highly lobed if floating), with expanded, sheathing stalks. Flowers are bisexual, emersed above water, borne 1-few on a stout stalk, with 5–6, shiny, yellow petals (7–12 mm long), and green centers. Fruit small (2.5–3.5 mm long), smooth, with a winged margin, and beak (1–2 mm long).

Habitat & Distribution

Ranunculus flabellaris is a native species of open, shallow (30–100 cm deep), acidic or alkaline, waters of ponds, bogs, swamps, and streams. It is widespread in CT, but rare and state-listed as a ***concern*** in RI.

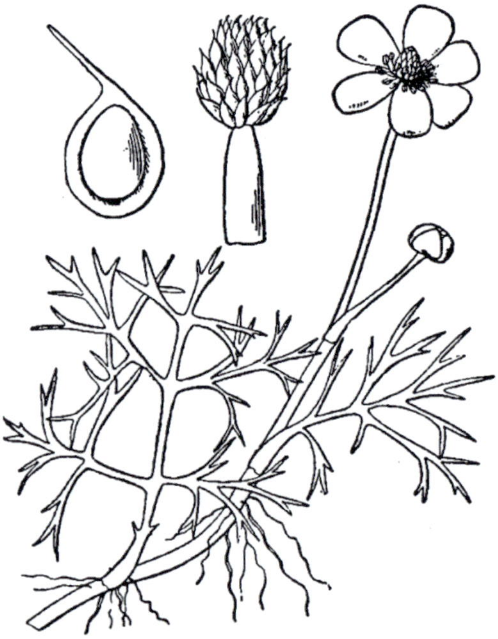

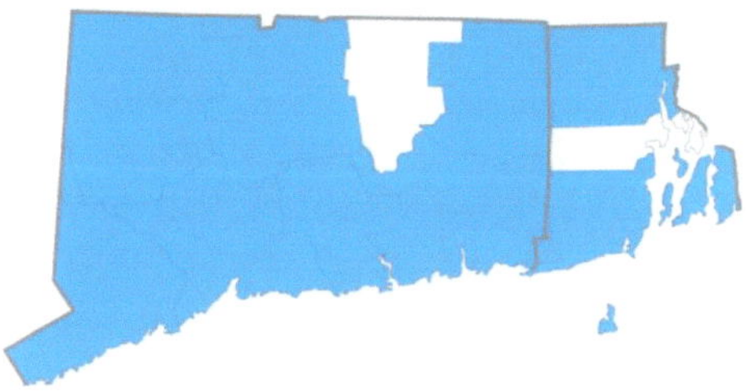

CURSED CROWFOOT *Ranunculus sceleratus* L.

Cursed crowfoot is a submersed annual with floating hollow stems rooted at base. Leaves are all basal (arising from submersed base of plant), with long stalks, and floating, deeply lobed blades (divided into 3–5 main segments). Flowers are bisexual, emersed above water on branching clusters, with five, small (2–5 mm long), yellow petals, and green centers. Fruit small (1.5–2 mm long), transversely finely wrinkled, with a minute beak.

HABITAT & DISTRIBUTION

Ranunculus sceleratus is a presumed native of shallow, fresh to brackish waters of marshes, swamps, and tidal marshes. It is uncommon in CT and rare and state-listed as a ***concern*** in RI.

Note: This species is highly variable and adaptable. Described here is the submersed form of the species. More often, plants will be erect, largely emergent, with alternate leaves supported on a hollow stem.

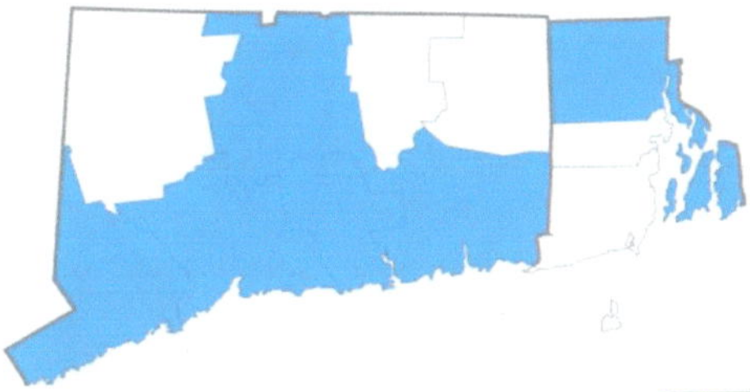

Ruppia — Ruppiaceae
Wigeon-grass

WIGEON-GRASS *Ruppia maritima* L.

Wigeon-grass, or Beaked ditch-grass, is a delicate, rooted, entirely submersed, marine perennial with rhizomes and thread-like stems (<1 mm wide). Leaves are alternate, thread-like (6–10 cm long, <0.5 mm wide), green, blunt-tipped, and fused to translucent stipules which sheath the stems at nodes. Flowers are bisexual, underwater, borne in pairs within a leaf sheath when young. Fruits ovoid (2 mm long), fleshy, short-beaked, each borne on a long stalk (1–2 cm long), in clusters of four at end of a stalk (<25 mm long) which can sometimes coil.

HABITAT & DISTRIBUTION

Ruppia maritima is a native, salt-tolerant plant of ponds, streams, ditches, and tidal pools. Common in shallow, brackish and saline waters of coastal CT and RI.

Note: Can be confused with Horned-pondweed (*Zannichellia palustris*).

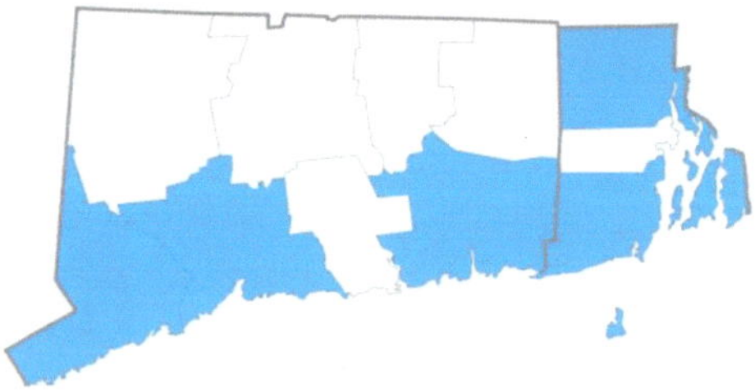

Sagittaria – Alismataceae

Arrowhead

The Arrowheads are rooted, submersed, floating-leaved, or emersed perennials, often with two leaf types on same plant. Leaves arise from a basal rosette and are highly variable in shape, influenced by water depth. Roots are septate and fibrous. Flowers are showy, insect-pollinated, with three white petals, and mostly unisexual (grouped on same plant), clustered in bracted whorls of three around a usually upright stem. The upper whorls are male flowers and lower female/fruiting (or bisexual). Flattened, winged, seed-like fruits (achenes) are crowded into a spherical cluster with persistent sepals at the base.

Seven truly aquatic species, all native, occur in CT and RI. Other Arrowheads (i.e., *S. latifolia* Willd. and *S. engelmanniana* J.G. Sm.) occupy shallow pond margins and wetlands in CT and RI. These species (excluded here), which can be very common, have firm, erect, foliage with obvious blades, commonly lobed at their bases.

KEY TO SPECIES

1. Sepals closely appressed in fruit, enclosing the fruit head; stalk of fruit head recurved........2
1. Sepals spreading or recurved in fruit, not enclosing the fruit head; stalk of fruit head ascending to spreading........4
   2. Leaves thick and spongy (5–10 mm wide), phyllodial; flowering stem emersed with 1–2 whorls of flowers, lower flowers bisexual........ *S. montevidensis*
   2. Leaves flattened or lens-shaped in cross-section (1–4 mm wide), phyllodial; flowering stem usually floating with 2–10 whorls of flowers, lower flowers female........3
3. Leaves flattened, ribbon-like, and flexuous (>30 cm long); flowering stems 10–200 cm long; plants often in swift freshwater rivers........*S. filiformis*
3. Leaves lens-shaped in cross-section, often firm (5–30 cm long); flowering stem 5–40 cm long; plants coastal, mostly in tidal brackish waters........*S. subulata*
   4. Stamens with filaments covered in short hairs; leaf blades (if present) without lobes........5
   4. Stamens with hairless filaments; leaf blades with basal lobes........*S. cuneata*
5. Leaves all phyllodial, essentially round in cross-section, generally tapering from near the base to tip........ *S. teres*
5. Leaves either phyllodial and flattened with parallel margins that taper somewhat abruptly to the apex, or with a flattened, expanded blade at tip........6
   6. Lowest whorl of female flowers borne on 1–3 cm long stalks; flowering stem straight, without a distinct bend; fruits 1.5–2 mm long........*S. graminea*
   6. Lowest whorl of female flowers stalkless or short-stalked (≤0.5 cm long); flowering stem often with a conspicuous bend at the lowest whorl of flowers; fruits 2.5–4 mm long........ .... *S. rigida*

S. graminea

S. teres *S. graminea*

S. cuneata

Northern arrowhead *Sagittaria cuneata* Sheldon

Northern arrowhead is a variable species with submerged, flattened (2.5–45 cm long, 1.5–11 cm wide), phyllodial leaves and floating (or emersed) leaves. Floating leaves have expanded blades (75–90 mm long, 35–40 mm wide), usually with basal lobes (shorter than rest of blade), and 3-sided leaf stalks. Flowers (to 25 mm wide) are emersed, with short petals (7–10 mm long), and recurved sepals that do not enclose the flower or fruit head. Stamens have linear, hairless filaments. Fruit (1.8–2.6 mm, 1.3–2.5 wide) is keeled with a diagnostic short (0.1–0.5 mm long), erect beak at the apex.

Habitat & Distribution

Sagittaria cuneata is a native, perennial species of calcareous pools, sluggish streams, and lakes. It is rare and state-listed as ***endangered*** in CT. Often occurs in open, deep waters (>1 m), and when flowing, plants take on a growth form with mostly long, broad, ribbon-like submersed leaves and relatively few floating leaves.

Notes: Similar to the widespread Common arrowhead (*S. latifolia*), but the submersed leaves, floating blades, shorter petals (≤10 mm), and ascending, short (≤5 mm) fruit beak of *S. cuneata* distinguish it. When plants are sterile, its ribbon-like leaves may be difficult to distinguish from *Vallisneria* and *Sparganium*.

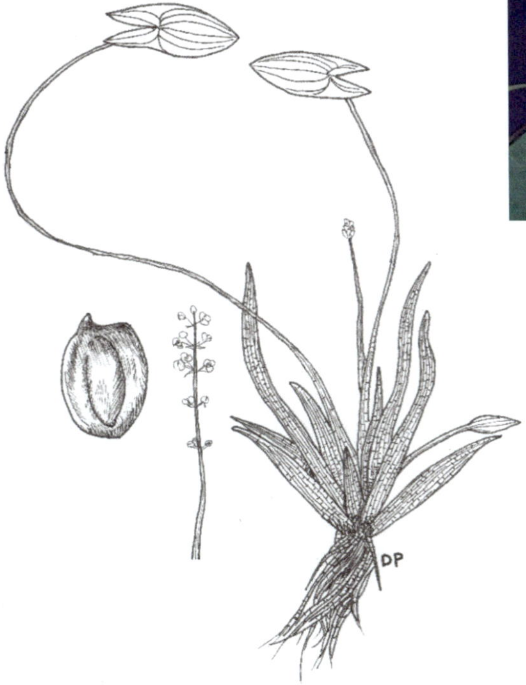

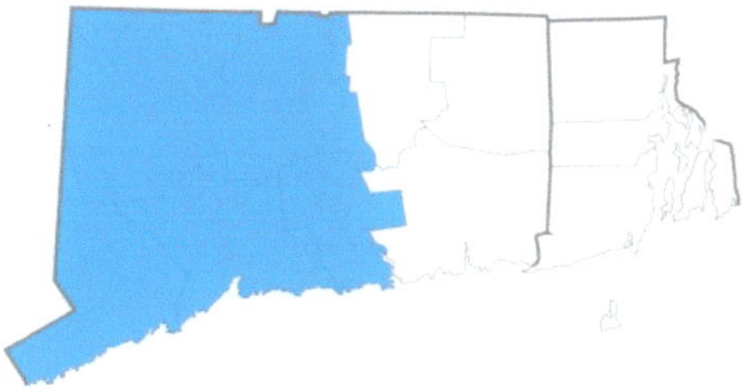

NARROW-LEAVED ARROWHEAD *Sagittaria filiformis* JG Sm.

Narrow-leaved arrowhead is a submerged, perennial, riverine plant. Leaves are submersed, ribbon-like, and long (30–250 cm long, 0.1–1.5 cm wide), or sometimes floating at ends with a dilated linear-ovate blade (to 0.5 cm wide). Flowers (to 30 mm wide) borne on a long (10–200 cm), usually floating, stem with 4–10 whorls of flowers. Sepals are erect and closely appressed in female flowers, enclosing the flower (or fruit head). Stamens have dilated, hairless filaments. Stalk of fruit head recurved,1.5–4.5 cm long. Fruits have an erect (1 mm) lateral beak.

HABITAT & DISTRIBUTION

Sagittaria filiformis is a native Coastal Plain species. It occupies shallow to deep reaches of slow to swift flowing waters, including tidal freshwaters. Frequent in CT and RI.

Notes: This species is not known to produce fruit in our area. Only plants of deep waters produce flowers. It is similar to Awl-leaved Arrowhead of brackish tidal shores of coastal CT which has shorter (≤30 cm), stiffer leaves.

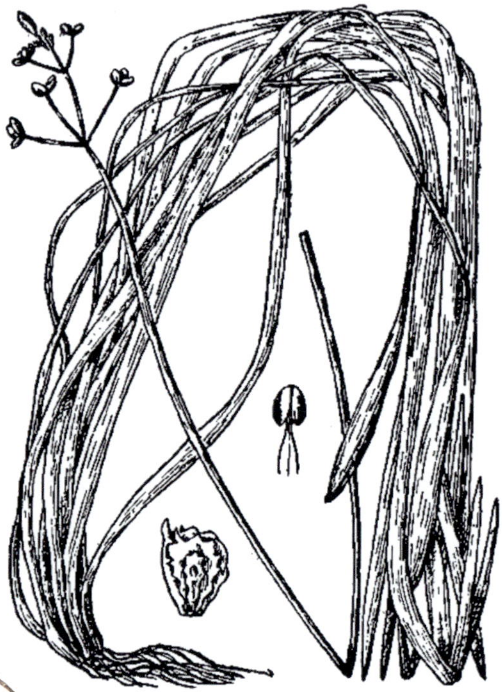

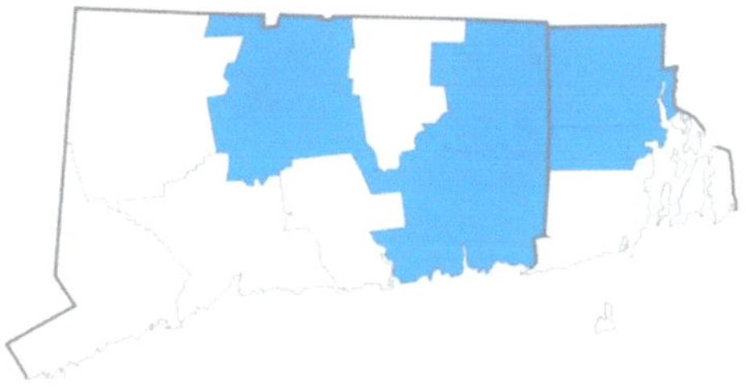

GRASS-LEAVED ARROWHEAD *Sagittaria graminea* Michx.

Grass-leaved arrowhead is a submerged (or emergent) rhizomatous perennial. Leaves are submersed, flattened, and phyllodial (6.5–35 cm long, 0.5–1 cm wide) with parallel margins that taper somewhat abruptly to the tip, or can have emersed leaves with a flattened, expanded blade (linear to linear-oval in shape, 2.5–17.5 cm long, 0.2–4 cm wide) lacking lobes. Flowers (to 23 mm wide) borne on an emergent, straight (without a distinct bend), stem with 1–12 whorls of flowers. Sepals are spreading or recurved in fruit, not enclosing the fruit head. Lowest whorl of female flowers borne on stalks (1–3 cm long). Stamens have dilated filaments covered in short hairs. Stalk of fruit head ascending to spreading. Fruits have an erect (0.2 mm) lateral beak.

HABITAT & DISTRIBUTION

Sagittaria graminea is a native species of ponds, rivers and freshwater tidal marshes. Widespread in CT but rare and listed as a species of ***concern*** in RI.
Note: In our region, it often occurs as submerged rosettes without flowers/fruit.

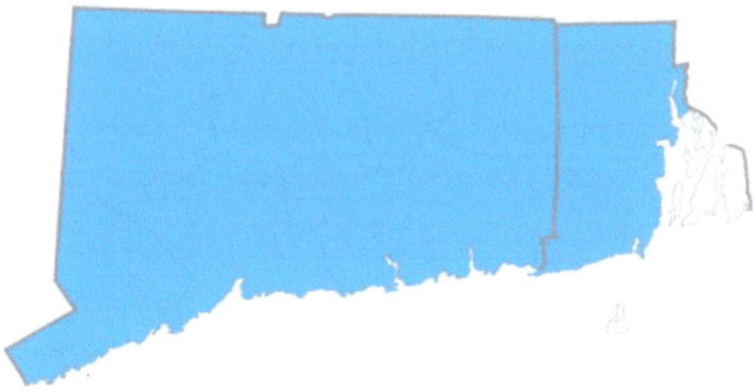

SPONGY-LEAVED ARROWHEAD *Sagittaria montevidensis* Cham. & Schlecht.

Spongy-leaved arrowhead, or Estuary arrowhead, is a submersed, dimunitive, annual plant. Leaves are submersed, thick, spongy, and strap-like (4–17 cm long, 0.5–1 cm wide). Leaves sometimes have expanded, spongy, lance-shaped blades. Flowers (ca. 20 mm wide) borne on an emersed stem, shorter than the leaves, singly or with 1–2 whorls of flowers. The lowest flowers are bisexual (with ring of sterile stamens). Sepals are erect and closely appressed in female flowers, enclosing the flower (or fruit head). Stamens have linear, hairless filaments. Stalk of fruit head thickened and recurved, 0.5–4.2 cm long. Fruits have a horizontal lateral beak (0.4–0.8 mm).

HABITAT & DISTRIBUTION

Sagittaria montevidensis is a native, coastal plant found on mudflats and shores of freshwater tidal rivers, typically exposed on mud flats at low tide. It is uncommon in CT.

Note: Our plants represent ssp. *spongiosa* (Engelm.) Bogin.

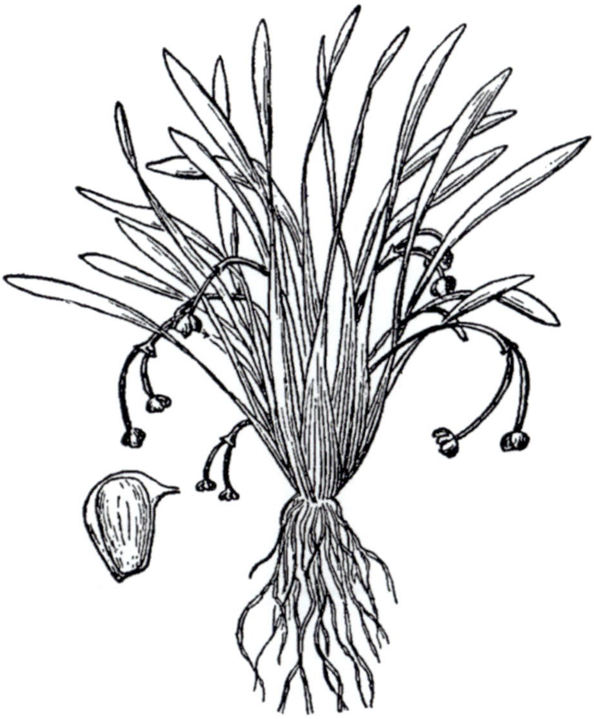

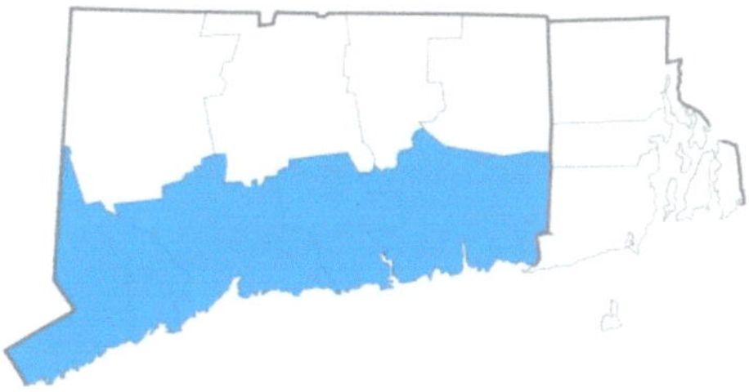

SESSILE-FRUITED ARROWHEAD *Sagittaria rigida* Pursh

Sessile-fruited arrowhead is a submerged (or emergent) perennial with slender stolons. Leaves are submersed, stiff, flattened, and phyllodial (30–70 cm long, 0.5–1 cm wide) with parallel margins that taper somewhat abruptly to the tip, or can be emersed with an expanded blade (linear to oval in shape, 5–15 cm long, 0.6–12 cm wide) commonly lacking lobes. Flowers (to 30 mm wide) borne on an emergent stem, conspicuously bent at the lowest whorl of flowers, with 2–8 whorls of flowers. Sepals are recurved and not enclosing the flower or fruit head. Lowest whorl of female flowers stalkless or borne on very short stalks (≤0.5 cm long). Stamens have dilated filaments covered in short hairs. Fruits have a recurved, lateral beak (0.8–1.4 mm).

HABITAT & DISTRIBUTION

Sagittaria rigida is a native species with highly variable leaves (influenced by water depth) which occurs in shallow to deep, fresh or brackish, waters of rivers, ponds, and pools. Plants are uncommon in CT. **Note**: When plants occur only as submersed, sterile rosettes it is impossible to distinguish from Grass-leaved arrowhead (*S. graminea*).

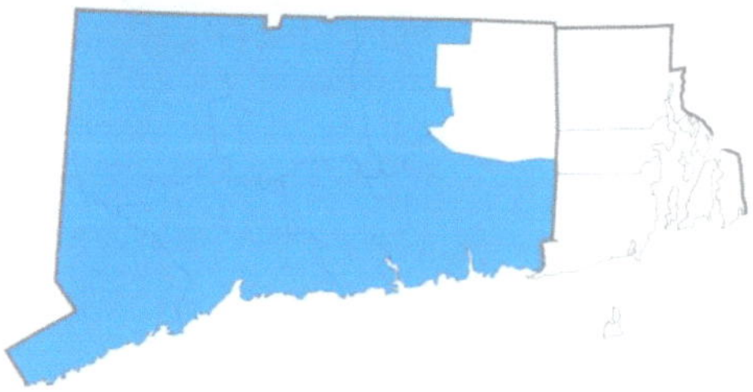

AWL-LEAVED ARROWHEAD *Sagittaria subulata* (L.) Buch.

Awl-leaved arrowhead is a submerged perennial, low-growing, tidal plant. Leaves are submersed, phyllodial (5–30 cm long, 0.1–0.4 cm wide), and lens-shaped in cross-section (i.e., two convex surfaces), sometimes with a linear-ovate blade (to 1.5 cm wide). Flowers (to 12 mm wide) borne on a floating stem (5–40 cm long), with 2–7 whorls of flowers. Sepals are erect and closely appressed in female flowers, enclosing the flower (or fruit head). Stamens have dilated, hairless filaments. Stalk of fruit head recurved. Fruits have an erect (0.4 mm) lateral beak.

HABITAT & DISTRIBUTION

Sagittaria subulata is a native coastal species, often submersed at high tide and exposed during low tides. It occupies shallow, estuarine brackish waters (salinity 0.5–18 ppt) and can be mat-forming. Rare and state-listed as a species of ***special concern*** CT (*State* ***historical*** in RI).

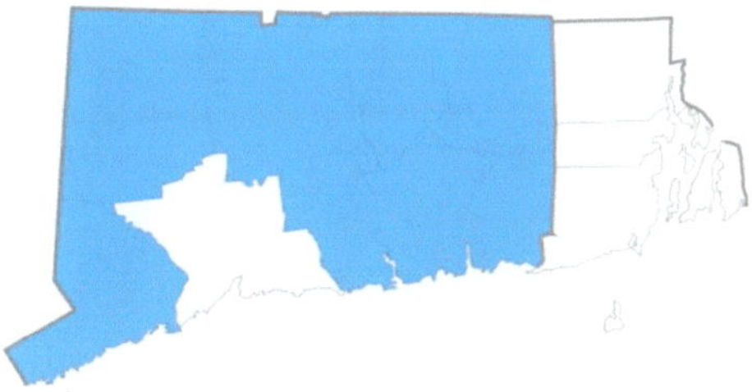

QUILL-LEAVED ARROWHEAD *Sagittaria teres* S. Wats.

Quill-leaved arrowhead is a submerged (or emergent) perennial with slender stolons. Leaves are submersed, rigid, linear (3.5–18.5 cm long, 0.15–0.4 cm wide), and essentially round in cross-section, thickest at base generally tapering to the tip, or can have longer leaves (to 60 cm) with emersed tips. Flowers (to 15 mm wide) borne on an emergent, straight (without a distinct bend), stem with 1–4 whorls of flowers. Sepals are spreading or recurved in fruit, not enclosing the fruit head; Lowest whorl of female flowers borne on stalks (1–3 cm long). Stamens have dilated filaments covered in short hairs. Stalk of fruit head ascending to spreading. Fruits have an erect (0.3–0.4 mm) to horizontal beak.

HABITAT & DISTRIBUTION

Sagittaria teres is a native Coastal Plain species of mostly sandy, acidic ponds. It is rare and state-listed as ***endangered*** in both CT and RI. Plants can be locally abundant and often inconspicuous as non-flowering, submersed rosettes.

Notes: Most similar to Grass-leaved arrowhead (*S. graminea*), especially when occurring as submersed basal rosettes, but leaves are always round in cross-section and never form blades.

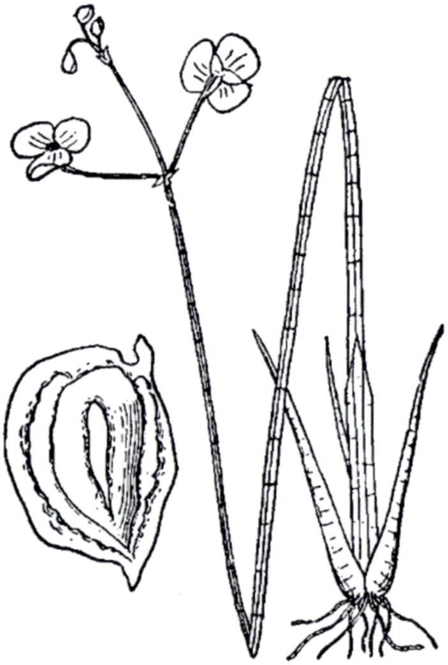

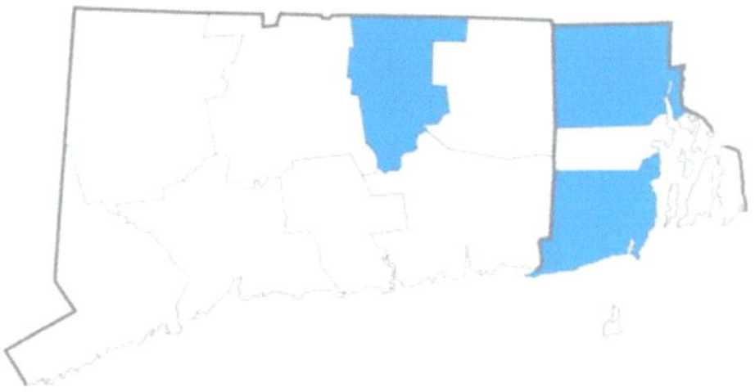

Schoenoplectus – Cyperaceae
Bulrush

WATER BULRUSH — *Schoenoplectus subterminalis* (Torr.) Soják.

Water bulrush is a rooted, grass-like, perennial, with submerged, flaccid, thread-like leaves clustered along slender (1 mm diameter), soft rhizomes and erect flowering stems. Leaves are alternate, submerged (or tips floating), 3–20, thin (to 1 mm wide), flexuous, with sheathing bases. Flowers are minute, borne in a small (4–5 mm long), straw-colored cluster at tip of a slender, upright stem (20–150 cm long), extending above water surface. A single, leaf-like (2–20 mm long) bract is attached below flower cluster, appearing like a continuation of the stem. Fruit is a seed-like (2.5 mm long), 3-sided achene.

HABITAT & DISTRIBUTION

Schoenoplectus subterminalis is a native species of lakes, bogs, and slow-moving streams. It is widespread in CT, but rare and state-listed as ***threatened*** in RI. This plant typically forms dense colonies and can colonize water to 1 m deep.

Notes: Sometimes leaves can be erect and C-shaped in cross-section. Similar to Robbins' spikesedge (*Eleocharis robbinsii*).

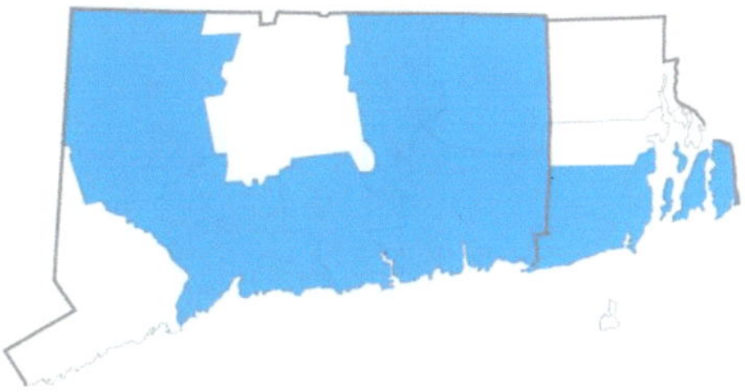

Sparganium — Typhaceae
Bur-reed

The Bur-reeds are rooted, perennials with buried rhizomes and long (10–50+ cm), floating stems. Leaves are submersed or partially floating, long (0.5–1+ m), linear (2–10 mm wide), flaccid, and generally flattened. Flowers are small, wind-pollinated, and unisexual (both on same plant), grouped into dense, spherical clusters (heads) held above water. Female flowers occupy the lower positions on the flowering stem and male flowers the uppermost positions. Fruits are a spherical aggregate of small, hard, beaked drupes.

Three aquatic species, all native and uncommon, occur in CT. Mature fruits are needed for correct identification. Several other Bur-reed species (e.g., *S. americanum* Nutt., *S. androcladum* (Engelm.) Morong, *S. emersum* Rehmann, and *S. eurycarpum* Engelm.) occur in the region that occupy wet shores or shallow waters. These emergent wetland species (excluded herein), however, are generally stout, with mostly stiff, erect, and emersed stems, and emersed, distinctly keeled leaves.

KEY TO SPECIES

1. Flowering stems with 1 cluster of male flowers; fruit clusters 5–12 mm in diam.; beak of fruit 0.5–1.5 mm long.. *S. natans*
1. Flowering stems with 3+ clusters of male flowers; fruit clusters 15–25 mm in diam.; beak of fruit 1.5–3.5 mm long..2
   2. Flowering stems branched; beak of mature fruit flattened and strongly curved; leaves 4–10 mm wide and flat..*S. fluctuans*
   2. Flowering stems unbranched; beak of mature fruit generally rounded and straight; leaves 2–5 mm wide with surface slightly convex.................................... *S. angustifolium*

S. angustifolium

NARROW-LEAVED BUR-REED *Sparganium angustifolium* Michx.

Narrow-leaved bur-reed is a submersed, rooted perennial. Leaves are submersed, with floating ends, long (to 1 m), ribbon-like (2–5 mm wide), with a slightly rounded surface. Flowering stems are unbranched with 2–5 female flower clusters, the lowest one on a stalk arising from the axil of a leafy bract. Male flower clusters number 4–10+. Fruit clusters 15–25 mm in diam. Fruits dull, reddish to brown (3–7 mm long) with a straight or slightly curved beak 1.5–2 mm long.

HABITAT & DISTRIBUTION

Sparganium angustifolium is a native species of shallow lakes and slow streams. It is uncommon in CT.

Note: Similar to the widespread Simple-stemmed bur-reed (*S. emersum*) which can sometimes have flaccid and floating leaves. Usually *S. emersum* has broader (6–12 mm wide), erect and emergent leaves which have a prominent keel along the lower surface of distal portions. Its fruit beak is also longer (2–6 mm).

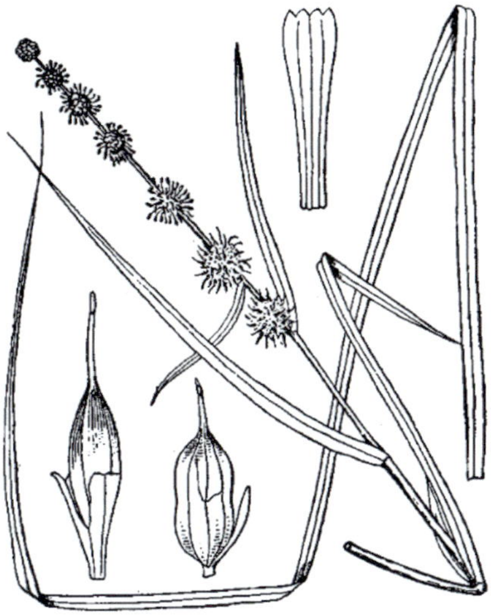

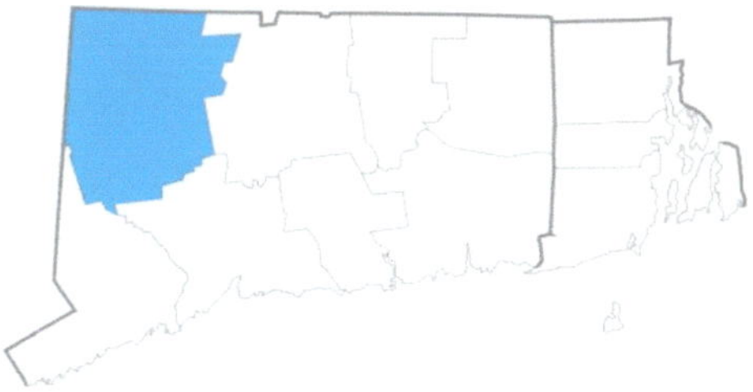

FLOATING BUR-REED *Sparganium fluctuans* (Morong) Robinson

Floating bur-reed is a submersed, rooted perennial. Leaves are submersed, with floating ends, strap-shaped (to 1.5 m long, 3–10 mm wide), and flat (but slightly keeled at base). Flowering stems are branching with 2–4 female flower clusters and 3–6 male flower clusters. Fruit clusters 15–23 mm in diam. Fruits dull, dark reddish-brown (3–5 mm long) with a flattened, strongly curved beak 2–3.5 mm long.

HABITAT & DISTRIBUTION

Sparganium fluctuans is a native species of shallow, still lakes or slow rivers. It is very rare and state-listed as ***endangered*** in CT.

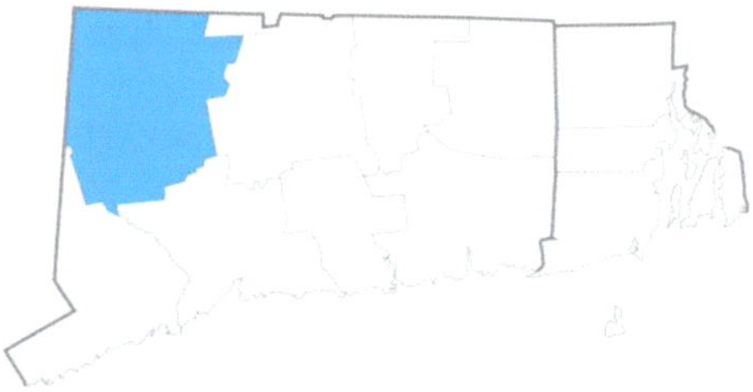

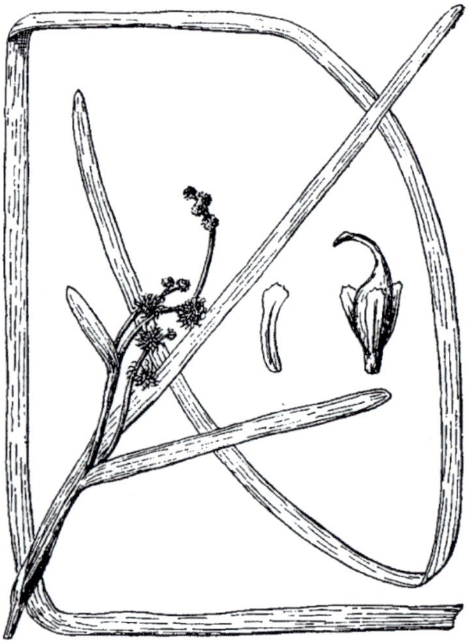

ARCTIC BUR-READ *Sparganium natans* L.

Arctic bur-reed is a submersed, rooted perennial. Leaves are submersed, with floating ends, strap-shaped (to 1.5 m long, 2–7 mm wide), and flat. Flowering stems are unbranched with 1–3 female flower clusters and only 1 male flower cluster. Fruit clusters 5–12 mm in diam. Fruits dull, greenish or brown (3–4 mm long) with a slender beak 0.5–1.5 mm long.

HABITAT & DISTRIBUTION

Sparganium natans is a native species of shallow lakes, peatlands, and slow streams It is very rare and state-listed as ***endangered*** in CT.

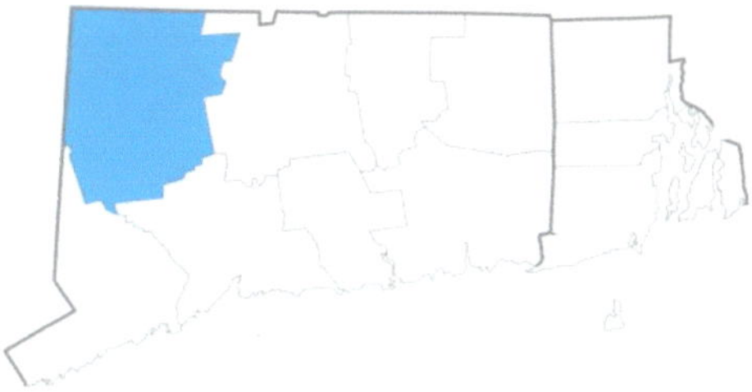

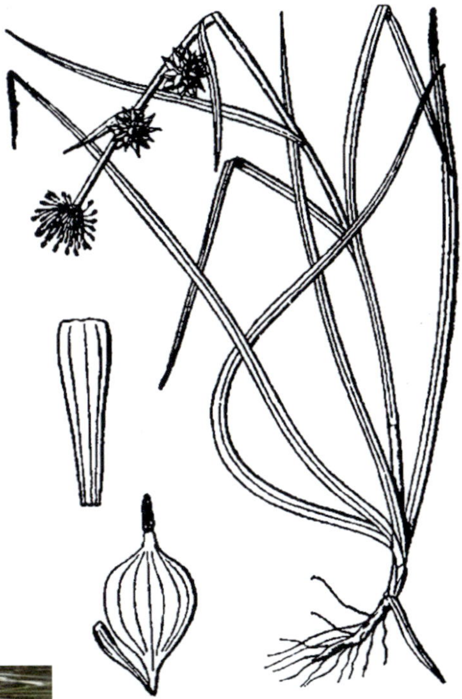

Spirodela – Araceae
Duck-meal

COMMON DUCK-MEAL *Spirodela polyrhiza* (L.) Schleid.

Common duck-meal is a diminutive, perennial, floating on water surface. Its body, or thallus (not differentiated into stem and leaf), is small (2–10 mm long, 1.3–10 mm wide), flattened, rounded to egg-shaped, with 5–21 obscure veins and often a purple dot on top surface, and solid red-purple beneath. Slender, 7–21, unbranched roots (to 30 mm long) are clustered on lower surface extending into water. Flowers and fruits are minute, rarely seen. Turions formed late in season.

HABITAT & DISTRIBUTION

Spirodela polyrhiza is a native species of ditches, ponds, lakes, and slow rivers. It is widespread in CT and RI.

Notes: Like most duckweeds, vegetative reproduction is extensive. Usually, multiple plants are connected and intermixed with other duckweeds (*Lemna* and *Wolffia*).

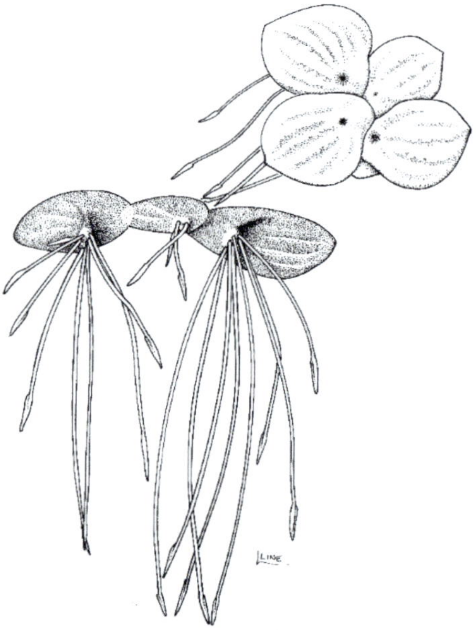

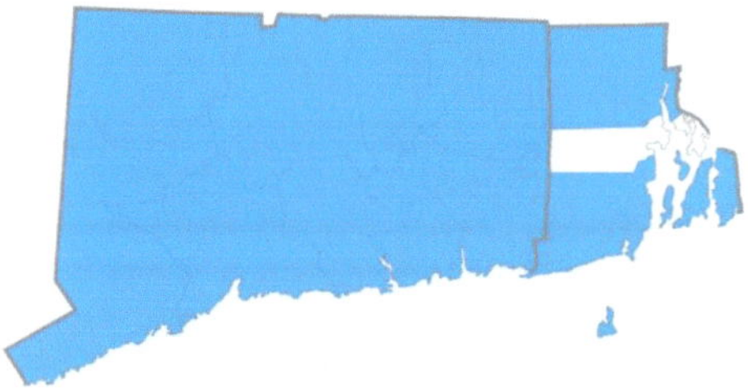

Stuckenia — Potamogetonaceae
False pondweed

SAGO FALSE PONDWEED *Stuckenia pectinata* (L.) Böerner

Sago false pondweed is a rooted, submersed, perennial with slender rhizomes which bear white tubers (1–1.5 cm). Upright stems are long (30–80 cm) and much-branched above. Leaves are alternate, all submersed, linear (3–15 cm long, 0.2–1 mm wide), opaque, channeled, 1-veined, stiff, and tapered to a sharp-pointed tip. Stipules (1–2 cm long) are attached to base of leaf for most of their length, forming an open sheath around the stem, the free tips extending as a small (0.8 mm long) ligule. Flowers are small and inconspicuous, clustered in an emersed, cylindrical spike (1–4 cm long) of 3–5 interrupted whorls of flowers, at end of a slender (3–12 cm), flexuous, submersed, stalk. Fruits are small (mostly 3–4.5 mm long), distinctly beaked, brown achenes with a round keel.

HABITAT & DISTRIBUTION

Stuckenia pectinata is a native plant of lakes, ponds, or streams. Fairly widespread in CT and RI. Colonies can form dense, bushy masses.
Note: The abundant, underground tubers of this plant are an important food for waterfowl.

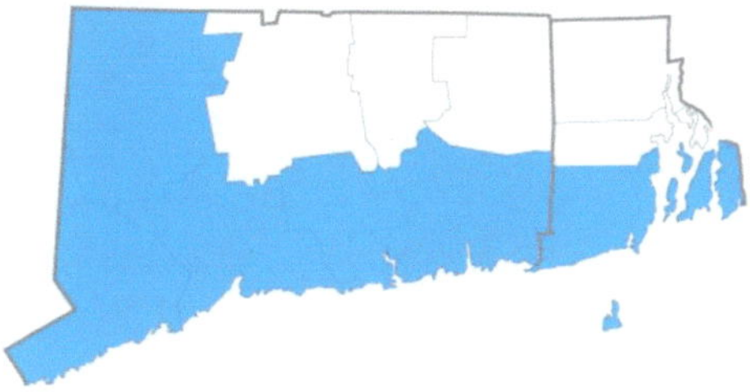

Trapa — Lythraceae
Water-chestnut

WATER-CHESTNUT *Trapa natans* L.

Water-chestnut, or Caltrop, is a rooted, annual plant with elongated (to 1 m), flexuous stems terminating with a floating rosette (to 50 cm diam.) of triangular leaves. Leaves are of two types: floating leaves are alternate, green (4–6 cm long, 4–8 cm wide), with triangular to diamond-shaped, broadly toothed blades, and swollen and spongy stalks. Submersed leaves are alternate or opposite, pinnately divided into linear (2–18 cm long), root-like segments. Flowers are bisexual, borne singly, above water, from a floating leaf axil, with 4 white petals. Fruit is dark, dry, nut-like (2–3 cm long, 2–4.5 cm wide) with 4 stout, spreading spines.

HABITAT & DISTRIBUTION

Trapa natans is a non-native species of lakes and margins of slow rivers. It is widespread but infrequent in CT or RI where it is **invasive** and banned. It prefers nutrient-rich, shallow (<2 m) waters and can reach high densities. This species is highly fecund, grows rapidly, and exhibits long seed dormancy. Floating rosettes can detach and float freely.

Notes: This is a highly distinctive plant with its inflated, buoyant, rosette of leaves and hard, black, spiny fruits.

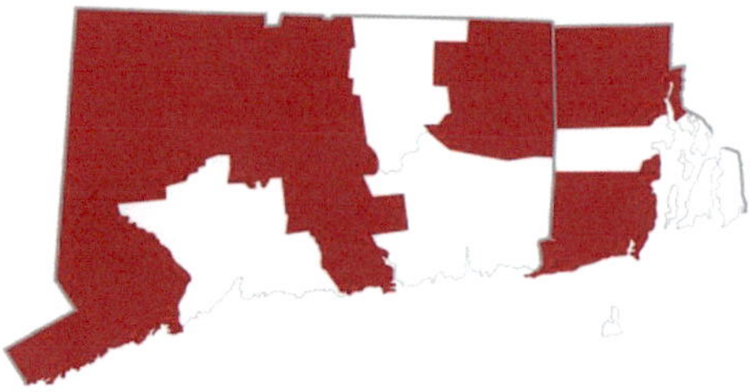

Utricularia — Lentibulariaceae
Bladderworts

The Bladderworts are delicate, mostly submersed, carnivorous perennials or annuals with narrow, rootless, green stems. Plants are mostly free-floating in water or in some cases anchored to a saturated substrate by buried branches. Leaves (actually leaf-like branches) are green, mostly divided into narrowly linear segments, and bearing tiny, sac-like, animal traps (i.e., bladders). Flowers are showy, yellow, or in a few species pink-purple, and borne singly or clustered at end of an elongated, unbranched stem elevated above water surface. Petals are asymmetrical, united, parted to form an upper and lower (larger) lip, and a spur (i.e., elongated hollow appendage). Fruits are rounded capsules. Many species form compact winter buds (turions) late in the season.

Twelve species occur in CT and RI, of which one is non-native. Without flowers present, proper species identification can be difficult.

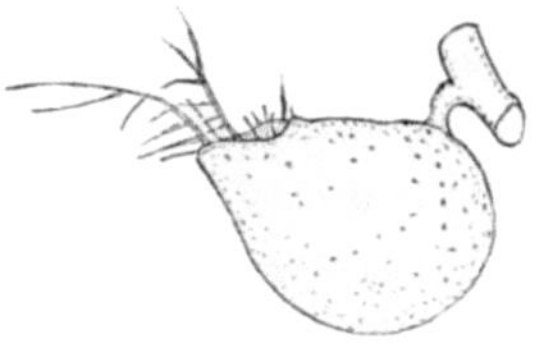

KEY TO SPECIES

1. Leaves few, unbranched and thread-like (or very narrowly linear), completely or mostly hidden in substrate; plants appearing terrestrial and only apparent when flowering; bladders few, 0.2–1.1 mm long……………… 2

1. Leaves numerous, finely forked 1-many times, all or mostly on stems floating under water, conspicuous; bladders well-developed, 1–5 mm long……………… 3

  2. Petals purple; one flower per stem; scale-like bracts on flowering stem paired and fused; leaves with ≥1 transverse partitions……………… *U. resupinata.*

  2. Petals yellow, 2–10 flowers per stem; scale-like bracts on flowering stem alternate; leaves without transverse partitions……………… 4

3. Flowering stem with a whorl of spongy, inflated branches near water surface functioning as floats……………… 5

3. Flowering stem without strongly inflated branches……………… 6

  4. Flower stalks 1–2 mm long; petal spur 7-14 mm long, pointing downward; fruits 4–5 mm Long……………… *U. cornuta*

  4. Flower stalks 4–15 mm long; petal spur 2.5–5 mm long, pressed against lower lip; fruits 1–1.5 mm long……………… *U. subulata*

5. Flowering stem 10–30 cm tall with 9–14 bright yellow flowers; floats 3–8 cm long and up to 8 mm wide, gradually tapered toward base……………… *U. inflata*

5. Flowering stem 3–5 cm tall with 1–7 dull yellow flowers; floats 1–4 cm long and 2–4 mm wide, with mostly parallel sides……………… *U. radiata*

  6. Leaves all whorled; bladders borne at tips of leaf segments and lacking external hairs; flowers pink-purple……………… *U. purpurea*

  6. Leaves mostly alternate or crowded at base; bladders scattered on sides of leaf

segments and with whisker-like hairs at the mouth; flowers yellow........................7
7. Leaf segments distinctly flat, with a visible midrib; floral bracts lobed at base................8
7. Leaf segments fine and hair-like, without a midrib; floral bracts not lobed........................9
8. Green leaf-like branches (above substrate) lacking bladders and whitish branches (buried in substrate) bearing bladders; leaf segments spine-toothed; bladders 1.5–4 mm long.. *U. intermedia*
8. Green leaf-like branches (above substrate) bearing some bladders; leaf segments not spine-toothed; bladders 1–2 mm.. *U. minor*
9. Plants small and simple, <15 cm long; principal leaves mostly forked 1–4 times, up to 1 cm long, the leaf segments without marginal spines; flowering stem 1–6 flowered..............10
9. Plants stout and bushy, 30–100 cm long; principal leaves mostly forked 5–17 times, 1–6.5 cm long, the leaf segments with minute marginal spines; flowering stem 2–14 flowered............11
10. All leaf-like branches green and bladder-bearing... *U. gibba*
10. Leaf-like branches of two types: those in water are green, with few bladders, and those buried in the substrate are whitish, bearing many bladders.............................. *U. striata*
11. Leaves 1.5–6.5 cm long, coarse, usually forked 6–17 times; bladders up to 6 mm long; flowering stem 6–14 flowered, with 1–3 scales below the lowest flowers; underwater flowers lacking; turions 5.5–13.5 mm thick.. *U. vulgaris*
11. Leaves 1–2 cm long, extremely delicate and limp, forked 5–9 times; bladders up to 2 mm long; flowering stem 2–8 flowered, without scales below the lowest flowers; bud-like, underwater flowers also borne singly on short, submersed stalks; turions 2–5 mm thick.. *U. geminiscapa*

Horned bladderwort is an inconspicuous perennial. Leaves are alternate, few, slender and unbranched (needle-like), and mostly hidden within the surface of a wet soil substrate. Leaves bear few, tiny (0.2–1.1 mm long) bladders also buried in the substrate. Flowers are bright yellow, very fragrant, each with a short stalk (1–2 mm long), a long (7–14 mm) spur pointed downward, and a lower lip 9–16 mm long. Flowers are crowded 2–6 on an upright, stout flowering stem (10–25 cm tall) with a few, widely spaced, alternate, scale-like bracts. Capsules are 4–5 mm long. No turions are produced.

HABITAT & DISTRIBUTION

Utricularia cornuta is a native species of bogs, pond shores, and river margins. It is widespread in CT and RI, more commonly of acidic waters.

Note: One of the so-called "terrestrial" species of bladderwort it is only apparent when flowering. Plants can be abundant, and form thin, tangled mats on a substrate which can dislodge and float.

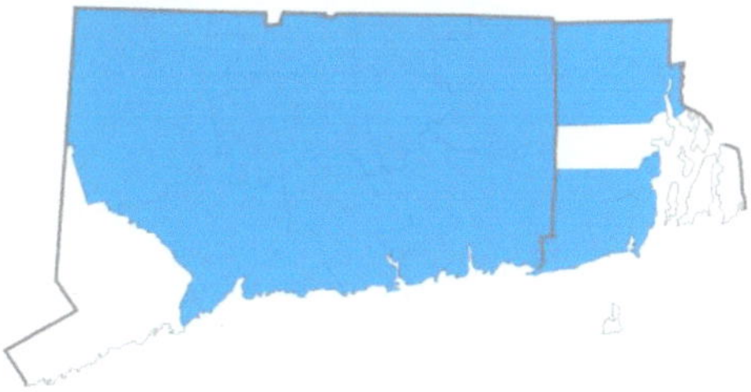

Mixed bladderwort *Utricularia geminiscapa* Benj.

Mixed bladderwort is a submersed, bushy, perennial with slender, elongated stems. Leaves are alternate, delicate (1–2 cm long), usually forked 5–9 times into limp, hair-like segments, with a few minute, marginal spines. Leaves bear numerous bladders (1–2 mm long). Flowers are yellow, with lower lip 5–8 mm long, and borne 2–8 on an upright flowering stem (5–15 cm tall). In late summer, bud-like, petal-less flowers are also borne singly at the tips of short (2.5–7.5 mm long), submersed stems. Turions are small (2–5 mm wide), loose, green, and lacking hairs.

Habitat & Distribution

Utricularia geminiscapa is a native species of, usually shallow, acidic, peatlands, pools, ponds, lakes, and streams. It is widespread in CT, but rare and state-listed as ***concern*** in RI.

Note: Similar to smaller plants of Common bladderwort (*U. vulgaris*) but overall reduced and finer, with submersed flowers, fewer emersed flowers, and smaller turions.

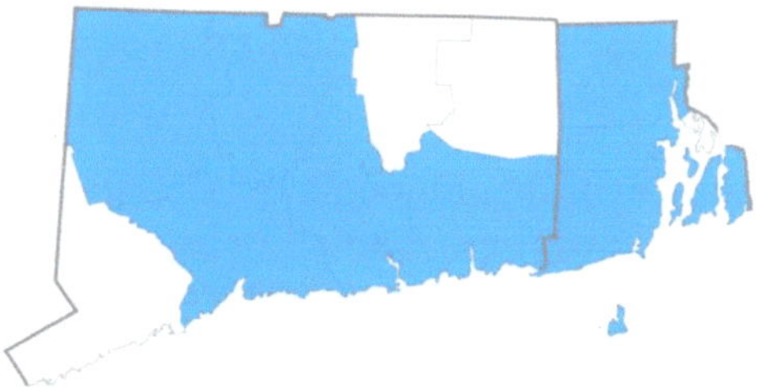

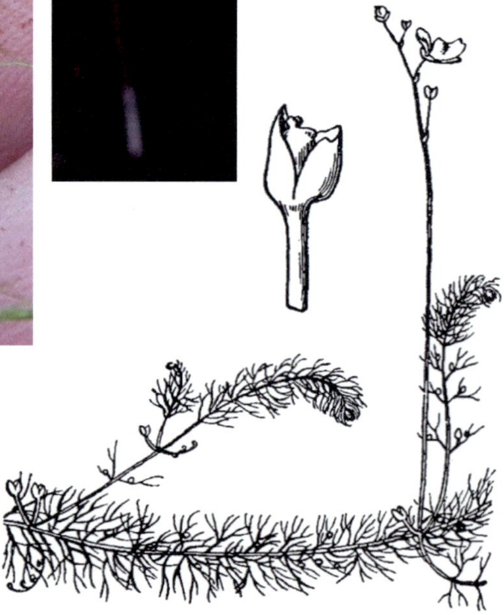

Creeping bladderwort *Utricularia gibba* L.

Creeping bladderwort is a delicate, submersed (or creeping), annual or perennial, with slender, short (to 10 cm long), intertwined, green stems. Leaves are alternate, short (rarely >5 mm long), mostly forked 1–2 times, and bear few, small (1 mm), bladders on sides. Flowers are yellow, with a blunt spur about half the length of the lower lip (5–6 mm long) and borne 1–2 on an upright flowering stem (5–10 cm tall). Capsules are 2–3 mm long. Turions are not produced.

Habitat & Distribution

Utricularia gibba is a native species of lakes and ponds. It is widespread and common in Ct, but rare and state-listed as ***concern*** in Ri.

Notes: Commonly found as free-floating, intricately tangled, masses, but flowering is usually limited to plants in very shallow water (2–3 cm deep) or those stranded on wet, exposed shores. Most similar to the uncommon Striped bladderwort (*U. striata*) which is distinguished by larger flowers and two types of branches.

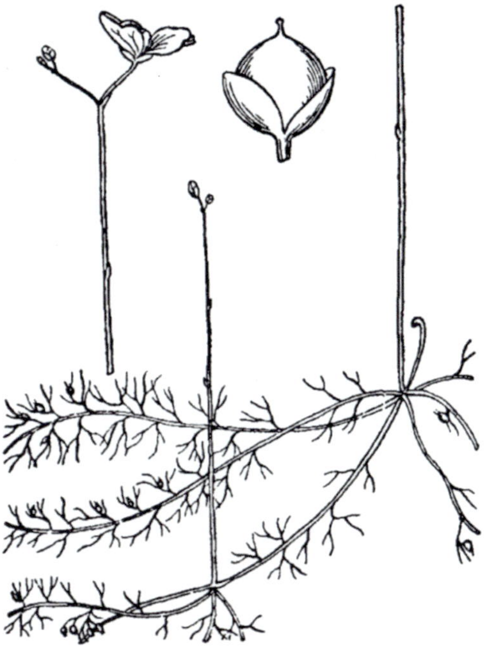

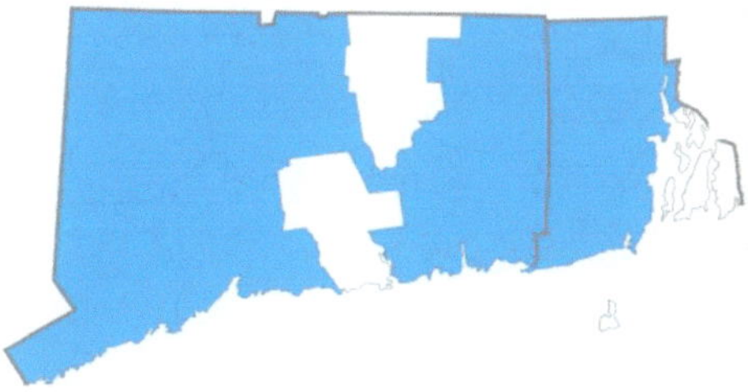

Swollen bladderwort *Utricularia inflata* Walt.

Swollen bladderwort is a submersed, perennial with slender, elongated stems. Submersed leaves (to 18 cm long) are alternate, bushy, repeatedly forked into thread-like segments bearing bladders on the sides. Uppermost leaves are whorled (5–10) around base of flowering stem and noticeably inflated, floating at water surface. Floats (3–8 cm long, to 8 mm wide) are spongy, gradually tapered toward flowering stem, with branching along half its length. Flowers are bright yellow, stalked (10–35 mm long), spur about half the length of the lower lip (20–25 mm long), and borne 9–14 (usually 10–11) on an upright flowering stem (10–30 cm tall).

Habitat & Distribution

Utricularia inflata is a non-native species of shallow, clear, lakes and ponds. It is **introduced** and uncommon into Ct and Ri. This species can reproduce vegetatively by stem fragmentation and tuber formation and has potential to become weedy.

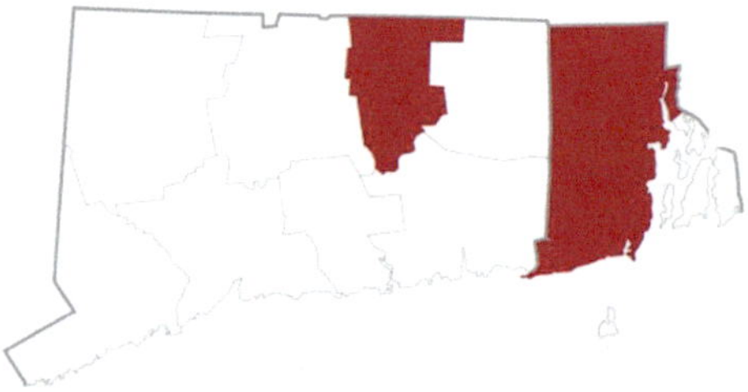

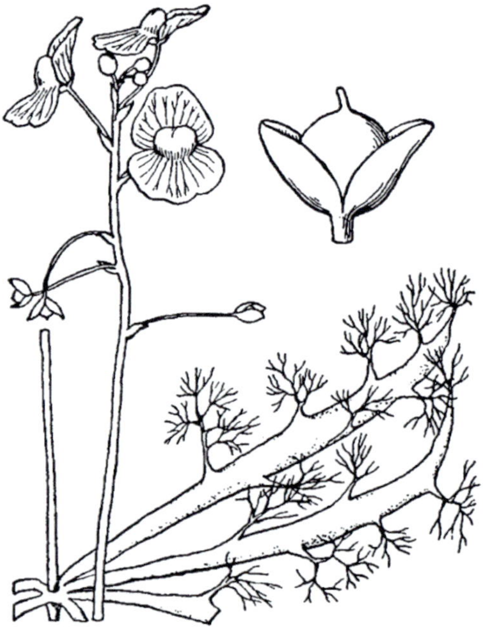

FLAT-LEAVED BLADDERWORT *Utricularia intermedia* Hayne

Flat-leaved bladderwort is a delicate, submersed (or creeping on substrate under water), perennial with slender stems. Leaves (above substrate) are alternate, green, fan-like, forked into strongly flattened, minutely spine-toothed, segments (5–20 mm long), lacking any bladders. Separate, whitish, leaf-like shoots (usually embedded in substrate) bear bladders (1.5–4 mm long). Flowers are bright yellow, with a long (8–12 mm) lower lip, well-developed spur (to 12 mm), and borne 2–5 on an upright flowering stem (6–20 cm tall). Turions are oval (3–10 mm long).

HABITAT & DISTRIBUTION

Utricularia intermedia is a native species of shallow waters of ponds, lakes, bogs, and marshes. It is widespread in CT, but rare and state-listed as ***concern*** in RI.

Notes: The species is intolerant of eutrophication. Similar to Lesser bladderwort (*U. minor*), but with green bladder-less leaves and larger, bright yellow flowers.

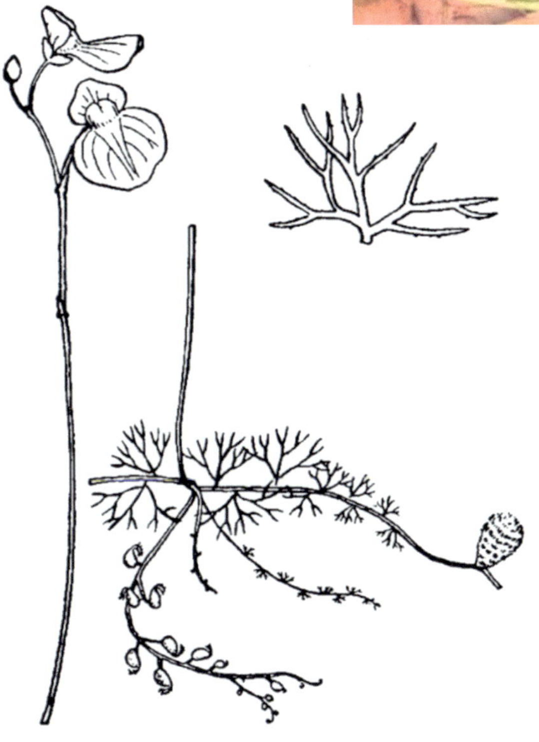

Lesser bladderwort *Utricularia minor* L.

Lesser bladderwort is a delicate, submersed (or creeping on wet substrate), perennial with slender stems. Leaves are alternate, numerous, fan-like, and usually forked 3–4 times into distinctly flattened segments (3–10 mm long). Green leaves bear few, small (1–2 mm long) bladders while whitish leaves embedded in substrate bear most bladders. Flowers are pale yellow, small (petals 6–8 mm long), with an obscure (almost absent) spur, and borne 2–8 on an upright flowering stem (4–15 cm tall). Turions are tiny (<3 mm), reddish-green, and rounded.

Habitat & Distribution

Utricularia minor is a diminutive, native species of shallow (<0.5 m deep), typically nutrient-poor, water of ponds, lakes, and rivers. It is widespread but uncommon in CT (state-listed as ***historical*** in RI).

Notes: Plants are rarely found in flower, so they can be easily overlooked. Flattened aspect of leaves easily seen under hand lens. Similar to Flat-leaved bladderwort (*U. intermedia*), but with the green leaves bearing bladders and smaller, pale flowers.

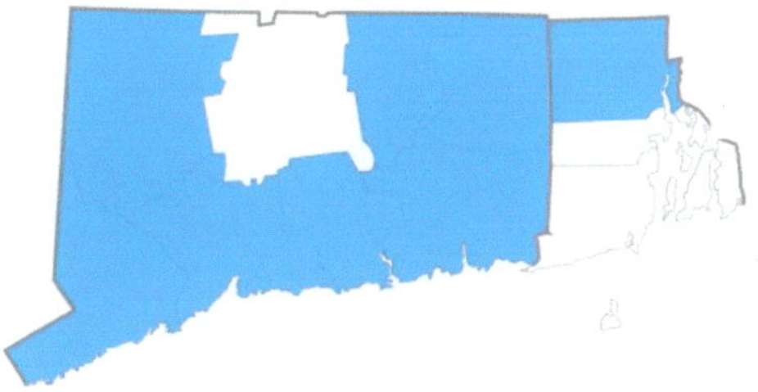

Purple bladderwort — *Utricularia purpurea* Walt.

Purple bladderwort is a submersed, perennial, with slender, elongated (to 1 m), much-branched stems. Leaves are in whorls of 5–7, conspicuous, stalked (1–1.5 cm long), forked several times into thread-like segments that bear bladders at their tips. Bladders are without external guide hairs. Flowers are pink-purple, with a blunt spur (2–5 mm long) pressed against the yellow-blotched lower lip, and borne 2–5 on an upright flowering stem (5–15 cm tall). Turions are not produced.

Habitat & Distribution

Utricularia pupurea is a native species of lakes, ponds, and slow streams. It is widespread and common in CT and RI.

Notes: Plants are free-floating and when conditions are optimal often become abundant. This species is found mostly in acidic, low nutrient waters and considered a good indicator species of phosphorous-poor water bodies.

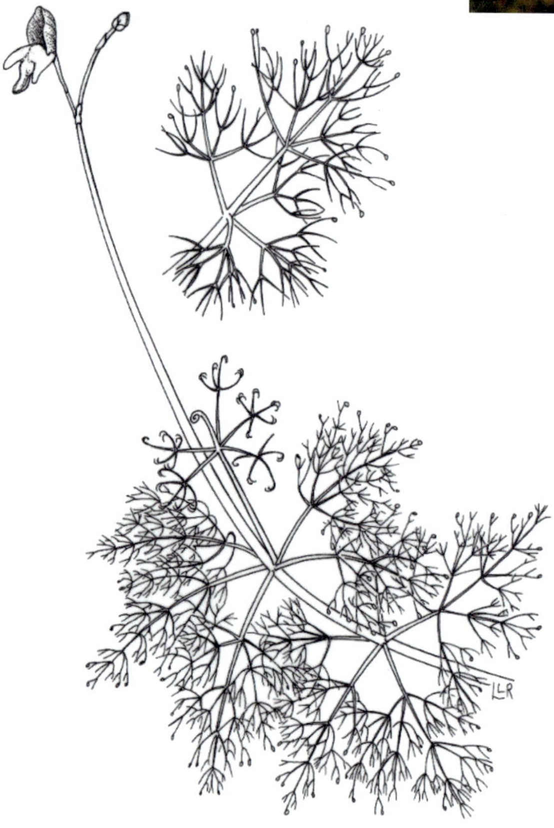

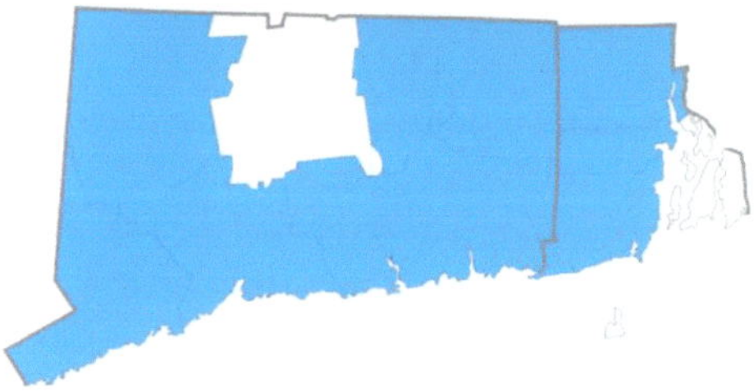

FLOATING BLADDERWORT *Utricularia radiata* Small

Floating bladderwort is a submersed, annual with slender, elongated stems. Submersed leaves (10–40 mm long) are alternate, repeatedly forked into thread-like segments, and bear many bladders on the sides. Uppermost leaves are whorled (4–7) around base of flowering stem and noticeably inflated, floating at water surface. Floats (1–4 cm long, 2–4 mm wide) are spongy, mostly parallel-sided, with branching limited to tips. Flowers are dull yellow, stalked (10–20 mm long), spur pressed against the lower lip (8–10 mm long), and borne 1–7 (mostly 3–4) on an upright flowering stem (3–5 cm tall). Turions are small (1 mm wide), roundish, and green.

HABITAT & DISTRIBUTION

Utricularia radiata is a native species of shallow, mostly acidic, ponds and lakes. It is widespread in CT and RI.

Notes: Similar to the more robust and bushier Swollen bladderwort (*U. inflata*) but with less densely branched leaves, smaller floats, shorter flowering stems, and fewer flowers. Difficult to identify without floats and flowers.

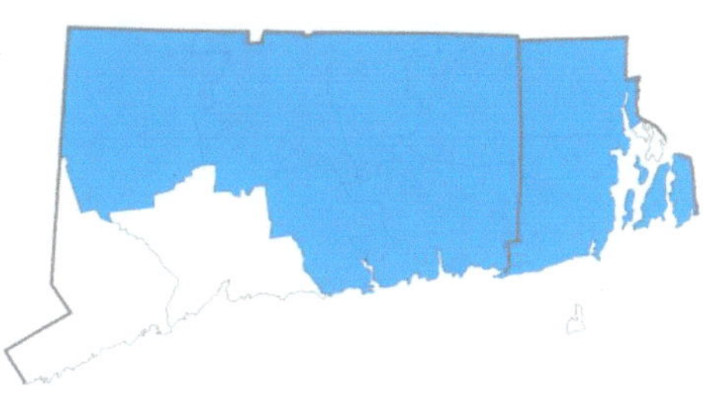

RESUPINATE BLADDERWORT *Utricularia resupinata* B.D. Greene

Resupinate bladderwort is an inconspicuous, delicate, perennial with horizontal, whitish, stems just below soil surface. Leaves are alternate, needle-like, unbranched, up to 3 cm long, with several transverse partitions. Leaves bear few, tiny (0.2–1.1 mm long) bladders often buried in the substrate. Flowers (1 cm) are purple, tipped backwards and facing upward, and borne singly at tip of an upright flowering stem (2–10 cm tall) with a pair of scale-like bracts fused around the flowering stem. No turions are produced.

HABITAT & DISTRIBUTION

Utricularia resupinata is a native species occurring usually in sandy, very shallow (<15 cm), nutrient-poor ponds or lakes. It is rare in CT and RI where it is state-listed as ***endangered*** and ***concern***, respectively.

Note: One of the so-called "terrestrial" species of bladderwort, it is only apparent when flowering, which is sporadic in our region. It can form tangled mats in deep water which can dislodge and float.

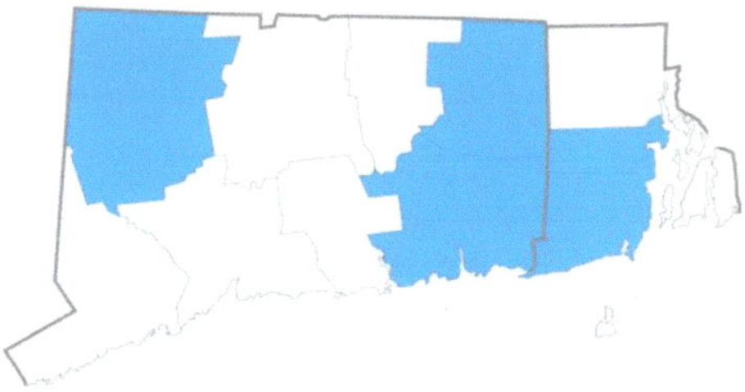

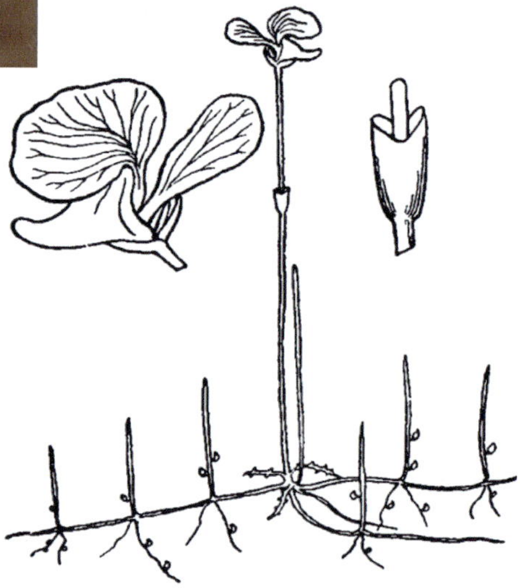

STRIPED BLADDERWORT *Utricularia striata* Leconte ex Torr.

Striped bladderwort is a delicate, submersed (or creeping), perennial, with slender, intertwined, green stems radiating from the flowering stem. Above the substrate, leaf-like branches are green, forked 3 or 4 times, and bear few, if any, small (1–2.5 mm long), bladders on sides. Below the substrate, branches are whitish, forked 1–4 times, and bear many bladders. Flowers are pale yellow, with a spur about ¾ the length of the lower lip (8–10 mm long) and borne 1–6 on an upright flowering stem (15–30 cm tall). Capsules are 2–3 mm wide. Turions are not produced.

HABITAT & DISTRIBUTION

Utricularia striata is a native species of acidic ponds. It is at its northern limits in CT and very uncommon.

Notes: It typically occurs in very shallow water (<10 cm) entangled in loose debris, but sometimes can be found in free-floating mats elsewhere. Similar to the more common Creeping bladderwort (*U. gibba*).

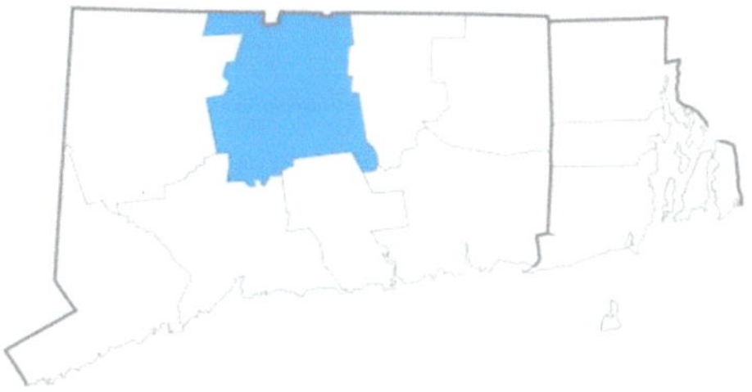

SLENDER BLADDERWORT *Utricularia subulata* L.

Slender bladderwort is an inconspicuous, delicate, annual species. Leaves (to 1 cm long) are alternate, slender, strap-like, unbranched, and mostly hidden within the surface of wet soil substrate. Leaves and stem branches can bear few, tiny (0.5 mm long), bladders also buried in the substrate. Flowers are yellow, each with an obvious stalk (4–15 mm long) and spur (2.5–5 mm) pressed against the lower lip (4–7 mm long). Flowers are borne 2–8 on an upright, zig-zagged, flowering stem (3–20 cm tall) with a few, alternate, scale-like bracts. Sometimes plants have smaller, whitish, closed, self-pollinating flowers. Capsules are 1–1.5 mm long.

HABITAT & DISTRIBUTION

Utricularia subulata is a native, coastal plain species occurring on sandy, or peaty sands, of acidic, pond shores or bogs. It is rare in RI where it is state-listed as of **concern**. One of the so-called "terrestrial" species of bladderwort, it is only apparent when flowering.

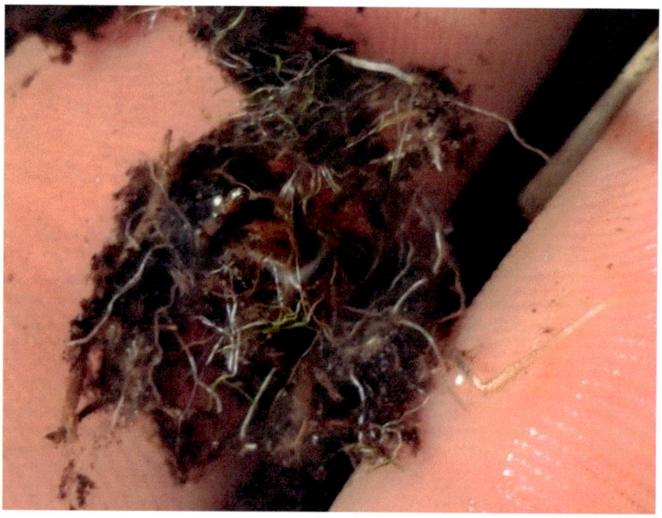

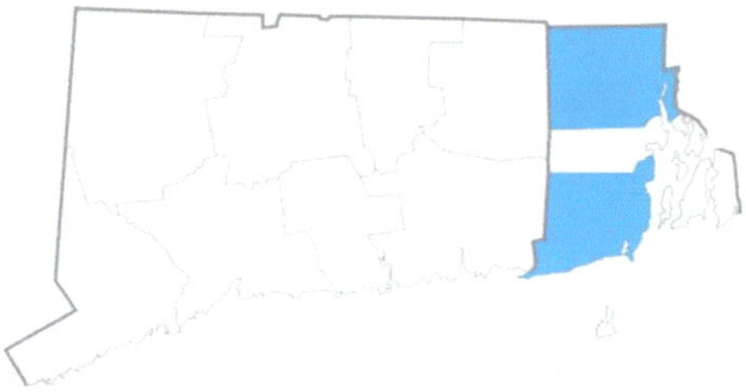

Common bladderwort *Utricularia vulgaris* L.

Common bladderwort, or Greater bladderwort, is a submersed, bushy, perennial with slender, elongated stems (30–100 cm long). Leaves are alternate, numerous, course (1.5–6.5 cm long), forked usually 6–17 times (and slightly zigzagged) into narrow segments, with minute marginal spines. Leaves bear numerous, conspicuous, bladders (1–6 mm long). Flowers are yellow, spur tip curved upward, lower lip 1–2 cm long, and borne 6–14 on an upright flowering stem (6–20 cm tall). Turions are large (1–2 cm long, 0.5–1.5 cm wide), rounded or elongated, abundantly haired, often mucilaginous, and green.

Habitat & Distribution

Utricularia vulgaris (= *U. macrorhiza* LeConte) is a native species of ponds, lakes, and streams. It is common and widespread in CT and RI.

Notes: This species, suspended and free-floating in water, has broad ecological tolerances (e.g., indifferent to water chemistry, flow, and depth) and is usually abundant. After pollination, flowering stems recurve and bend so fruit development takes place under water.

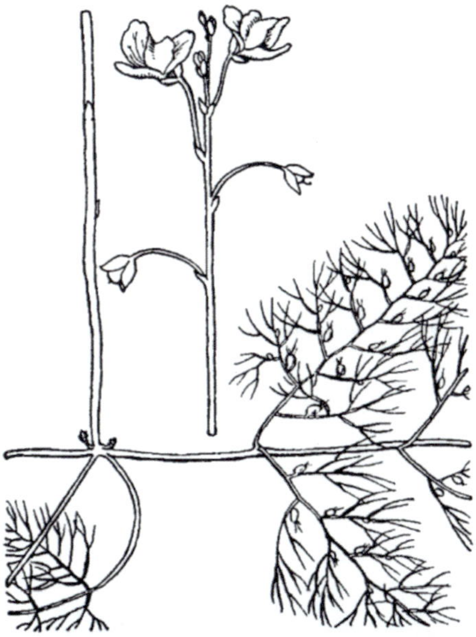

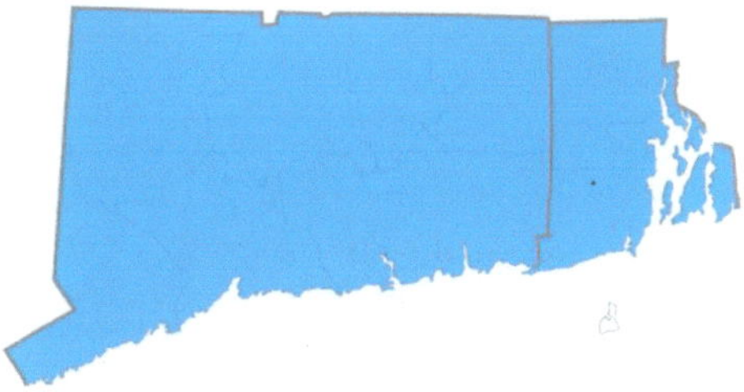

Vallisneria — Hydrocharitaceae
Tape-grass

AMERICAN EEL-GRASS *Valisneria americana* Michx.

American eel-grass, or Water-celery, is a rooted, perennial herb with long, submersed, ribbon-like leaves clustered along stolons. Leaves are basal, submersed (upper portions may be floating), flat, long (10–110 cm long, 0.5–1.5 cm wide), bluntly pointed, bright green, and somewhat translucent. The leaf midvein has 4–5 rows of lacunae on each side, giving the blade a 3-zone appearance (light-colored middle zone bordered by darker zone on sides). Flowers are unisexual (on separate plants) and small (2 mm wide) with transparent petals. Fruit is a cylindrical (50–120 mm long) capsule borne on end of an elongated, coiled stalk.

HABITAT & DISTRIBUTION

Vallisneria americana is a native species of fresh to brackish slow rivers, streams, or lakes. It is common and widespread in CT and RI. It is a colony-forming plant and a valuable waterfowl food. **Notes:** Numerous male flowers detach from the plant and float to surface while single female flowers are projected to surface tethered by a long stalk. After pollination, the stalk coils and pulls the fruit underwater. It can appear very similar to non-flowering plants of *Sparganium* in deep waters.

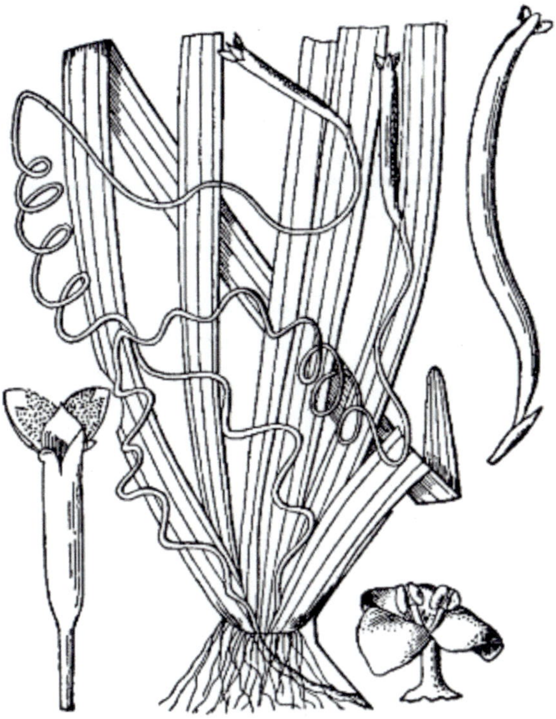

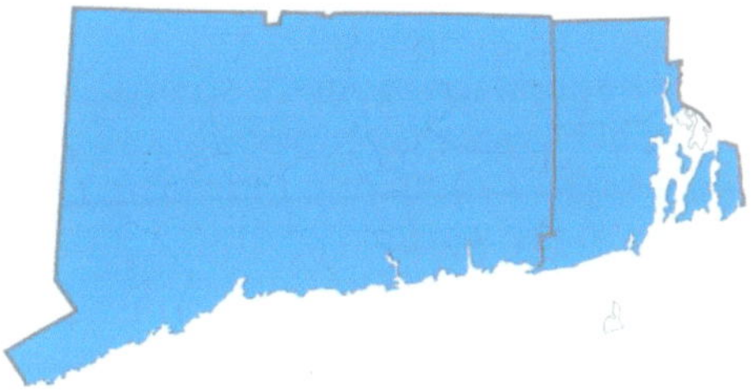

Wolffia — Araceae

Water-meal

The Water-meals are tiny (<1.5 mm long), green, rootless, perennials floating on surface of quiet waters. Plant body (called a thallus) is greatly reduced (not differentiated into stem and leaf), spherical, oval, or boat-shaped (i.e., flattened above and rounded below). Usually, two plants are connected. Flower and fruit are microscopic, produced on upper surface, and rarely seen.

Two species occur in CT and RI. Identification is best when plants are fresh. Individuals reproduce vegetatively quickly and are usually very numerous, often intermixed with other duckweeds (*Lemna* and *Spirodela*). This genus contains the smallest known angiosperm species in the world.

KEY TO SPECIES

1. Plants spherical, rounded in outline, largest individuals 0.5–0.8 (–1.2) mm wide .... *W. columbiana*
1. Plants oblong, longer than wide (shaped somewhat like segments of an orange), seldom over 0.5 mm wide by 1 mm long, upper surface flattened........................... *W. brasiliensis*

BRAZILIAN WATER-MEAL *Wolffia brasiliensis* Weddell

Brazilian water-meal is a tiny, rootless, free-floating perennial. Plants are oblong (0.5–1.6 mm long), longer than wide, and rounded at the ends. The upper surface is flattened marginally but raised in center with a prominent point.

HABITAT & DISTRIBUTION

Wolffia brasiliensis is a native plant of nutrient-rich, quiet ponds and sluggish rivers. It is sporadic in CT and RI.

Note: Very similar to Northern water-meal (*W. borealis* (Engelm.) Landolt) which may occur in CT. It is pointed at the ends with a minute upward bend, and the upper surface is flattened, smooth, without a central, raised point.

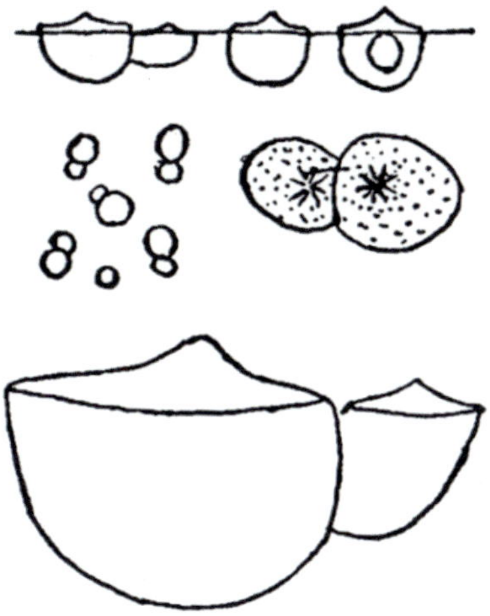

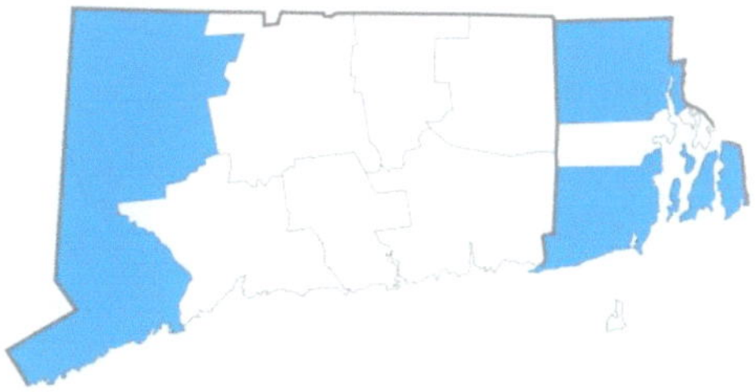

COLUMBIAN WATER-MEAL *Wolffia columbiana* Karst.

Columbian water-meal is a tiny, rootless, perennial, free-floating at water surface. Plants are spherical, rounded in outline. Largest individuals are 0.5–0.8 (–1.2) mm wide.

HABITAT & DISTRIBUTION

Wolffia columbiana is a native plant of nutrient-rich, quiet ponds and sluggish rivers. It is somewhat widespread in CT and RI.

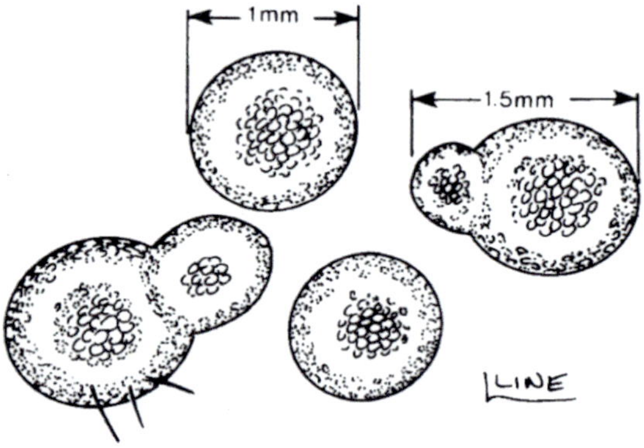

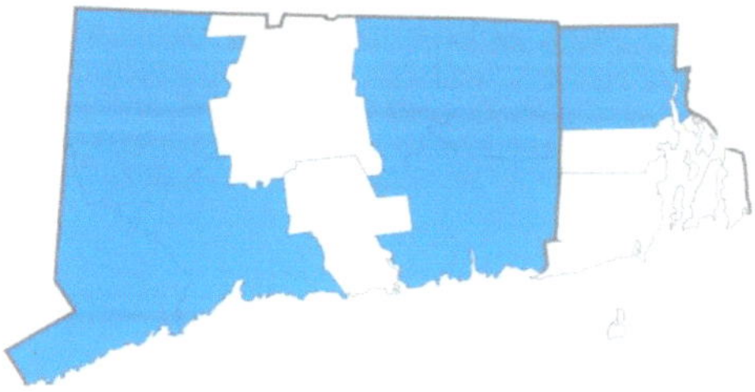

Zannichellia – Potamogetonaceae
Horned-pondweed

HORNED-PONDWEED *Zannichellia palustris* L.

Horned-pondweed is a delicate, rooted, annual plant, entirely submersed, with many, thin (0.5 mm wide) branching stems. Leaves are opposite (possibly obscured with smaller leaves in axils), green, linear (3.5–4 cm long, 0.2–1 mm wide) with sheathing stipules at their base, and pointed at tips. Flowers are unisexual and borne underwater in leaf axils. Fruit a cluster (2–5) of short-stalked (1 mm), flattened, achenes (2–3 mm long, 1 mm wide), slightly toothed on outer edge, with narrow extension (2 mm) at tip.

HABITAT & DISTRIBUTION

Zannichellia palustris is a native plant of freshwater lakes and tidal brackish waters. It is most common in hard waters near the coast but can be found more inland.

Note: Sometimes confused with Wigeon-grass (*Ruppia maritima*).

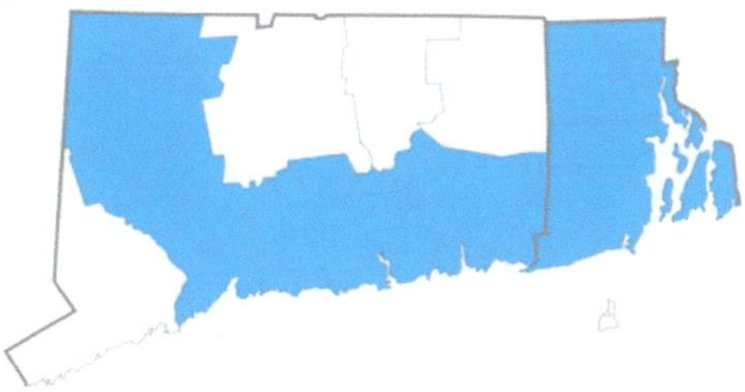

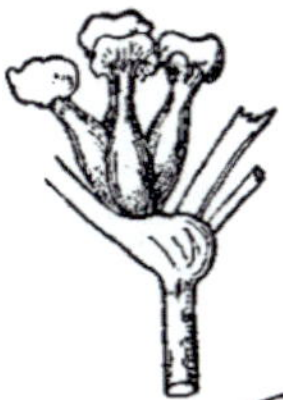

Zostera — Zosteraceae
Eel-grass

EEL-GRASS *Zostera marina* L.

Eel-grass, or Sea-wrack, is a grass-like, completely submersed, marine perennial with rooted, creeping rhizomes (2–6 mm thick) and flattened upright branches. Leaves are alternate, dark green, linear and ribbon-like (30–100 cm long, 2–12 mm wide), round-tipped, and sheathing the stem at base. Flowers are minute, submersed, unisexual (on same plant), lacking petals, in rows, enclosed within a leaf-like sheath (4–8 cm long) at end of a short (2–3 cm) stalk. Fruit a fleshy, oblong (2–5 mm long), beaked drupe.

HABITAT & DISTRIBUTION

Zostera marina is a native species most common in deeper subtidal (or less frequently in exposed intertidal areas) saline waters of sheltered ocean bays and inlets. It is widespread in coastal CT and RI. It can form vast, dense colonies or eel-grass "beds".

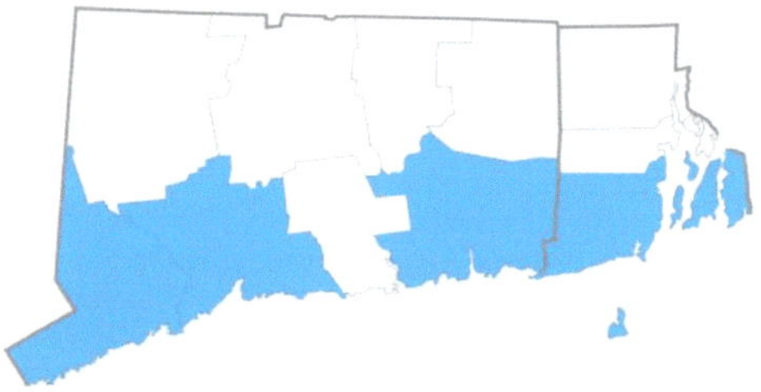

Notes: This is the only submersed flowering plant of marine waters in our region. As a sea grass, it is an extremely vital component of estuaries concerning sediment stabilization, nitrogen-fixation, animal food, shelter, and nurseries.

REFERENCES

Bell, M. 1985. The Face of Connecticut: People, Geology, and the Land. Bulletin 110. State Geological and Natural History Survey of Connecticut. Hartford, Connecticut.

Block, T.A. and A.F. Rhoads. 2011. Aquatic Plants of Pennsylvania: A Complete Reference Guide. University of Pennsylvania Press, Philadelphia.

Canavan IV, R.W. and P.A. Siver. 1994. Chemical and physical properties of Connecticut lakes, with emphasis on regional geology. Lake and Reserv. Manage. 10: 175-188.

Crow, G.E. and C.B. Hellquist. 2000. Aquatic and Wetland Plants of Northeastern North America, Vols. 1-2. University of Wisconsin Press, Madison.

Flora of North America Editorial Committee, eds. 1993+. Flora of North America North of Mexico. 21+ vols. New York and Oxford.

Gleason, H.A. and A. Cronquist. 1991. Manual of Vascular Plants of Northeastern United States and Adjacent Canada. New York Botanical Garden, Bronx, New York.

Haines, A. 2011. New England Wild Flower Society's Flora Novae Angliae. Yale University Press, New Haven, Connecticut.

Hutchinson, G.E. 1975. A Treatise on Limnology: Limnological Botany, Vol. 3. John Wiley & Sons, New York.

Les, D.H. 2017. Aquatic Dicotyledons of North America: Ecology, Life History, and Systematics. CRC Press, New York.

Les, D.H. 2020. Aquatic Monocotyledons of North America: Ecology, Life History, and Systematics. CRC Press, New York.

Mattson, M.D., P.J. Godfrey, M.F. Walk, P.A. Kerr, and O.T. Zajicek. 1992. Regional chemistry of lakes in Massachusetts. Water Resources Bull. 28: 1045-1056.

ILLUSTRATION & PHOTO CREDITS

Beyond my own photographs and illustrations, illustrations and photographs were utilized either from the public domain, with permission from authors, or according to various Creative Commons licenses. The vast majority of line drawings originated from Britton, N.L., and A. Brown. 1913. *An illustrated flora of the northern United States, Canada and the British Possessions. 3 vols.* Charles Scribner's Sons, New York (via USDA-NRCS PLANTS Database).

Photographs or drawings not listed with a credit below are either in the public domain or represent works of my own. For frequently cited sources, the following acronyms are used for brevity:

APP S.L. Winterton, *Aquarium and Pond Plants of the World*, Edition 3 (https://idtools.org/id/appw/)
CBH © C. Barre Hellquist; Used with permission.
MIN © Katy Chayka/Peter M. Dziuk, www.minnesotawildflowers.info. Used with permission.
UFL © Univ. Florida, Center for Aquatic and Invasive Plants. Used with permission.
WF USDA-NRCS PLANTS Database / USDA NRCS. *Wetland flora: Field office illustrated guide to plant species.* USDA Natural Resources Conservation Service.

Page 8, Figure 2, a–d: **UFL**
Page 9, Figure 3, a.: **UFL**; b.: **CBH**; d.: **UFL**; g.: **UFL**
Page 20, Drawing: **UFL**; Photos: K. Ozment, https://www.inaturalist.org/observations/18738845 (CCO-1.0); I. Taylar, https://commons.wikimedia.org/wiki/File:Azolla_caroliniana.jpg (CC-BY-2.0).
Page 22, Photos: R. Foster https://www.inaturalist.org/observations/92470784 (CC-BY-4.0); https://www.inaturalist.org/observations/31649871 (CC-BY-4.0).
Page 23, Photo: (leaf underside) **APP**
Page 24, Photos: L.J. Mehrhoff, https://commons.wikimedia.org/wiki/File:Cabomba_caroliniana_5447098.jpg (CC-BY-3.0); https://www.invasive.org/browse/detail.cfm?imgnum=5447111 (CC-BY-3.0).
Page 25, Drawing: **WF**
Page 26, Photos: (*C. palustris*) R. Durand, https://www.inaturalist.org/observations/68629881 (CC-BY-4.0); (*C. stagnalis*) dannali, https://www.inaturalist.org/observations/64844397 (CC0-1.0).
Page 27, Photo: M. Bowser, https://www.inaturalist.org/observations/15232773 (CC-BY-4.0).
Page 28, Photos: (habit) L. Allain, U.S. Geological Survey (Public Domain); (rosette) S. Kieschnick, https://www.inaturalist.org/observations/71669771 (CC-BY-4.0).
Page 29, Drawing: H. Coste and C. Flahault, *Flore descriptive et illustrée de la France, de la Corse et des contrées limitrophes*, 1901-1906 (Public Domain); Photo: R. Routledge, https://www.invasive.org/browse/detail.cfm?imgnum=5541152 (CC-BY-3.0)
Page 30, Drawing: H. Coste and C. Flahault, *Flore descriptive et illustrée de la France, de la Corse et des contrées limitrophes*, 1901-1906 (Public Domain); Photo: L.J. Mehrhoff, https://www.invasive.org/browse/detail.cfm?imgnum=5447134 (CC-BY-3.0)
Page 31, Drawing: **UFL**.
Page 32, Drawing: C.F. Reed, 1970. *Selected weeds of the United States*. USDA Agric. Res. Ser. Agric. Handbook 336 (Public Domain); Photo: (fruit) LacCore TMI website (http://lrc.geo.umn.edu/laccore/)
Page 33, Photo: (leaf) D. Suitor, https://www.inaturalist.org/observations/84200597 (CC-BY-3.0).
Page 35, Drawing: **UFL**; Photo: Lamiot, https://commons.wikimedia.org/w/index.php?curid=42528393 (CC-BY-SA 4.0).
Page 36, Drawing: **UFL**; Photo: (leaves) Amada44, https://commons.wikimedia.org/wiki/File:Eichhornia_crassipes_001.jpg (CC-BY-3.0); (flowers) C. Grenier, https://www.inaturalist.org/observations/21416617 (CC0-1.0).
Page 37, Photo: R. Foster, https://www.inaturalist.org/observations/21099340 (CC-BY-4.0).
Page 40, Photo: Bernt Knutsen, Norway
Page 41, Photo: G. Grzejszczak, https://www.inaturalist.org/observations/14359941 (CC-BY-4.0).
Page 43, Photo: N. Martineau, https://www.inaturalist.org/observations/12871902 (CC-BY-4.0).
Page 44, Photo: **MIN.**
Page 45, Drawings: **UFL**; Photo: R. Martin (Public Domain)
Page 46, Photo: C. Fischer, https://commons.wikimedia.org/wiki/File:ElodeaCanadensis.jpg (CC-BY-SA-3.0).

Page 47, Photo: C. Fischer**,** https://commons.wikimedia.org/wiki/File:ElodeaNuttallii2.jpg (CC-BY-SA-3.0).
Page 49, Photo: Jomegat, https://commons.wikimedia.org/wiki/File:Eriocaulon_aquaticum_6884.jpg (CC-BY-SA-3.0).
Page 51, Photos: (fruits) Connecticut Agricultural Experiment Station (Public Domain); (plants) C. Jacono, U.S. Geological Survey, Dept. of the Interior/USGS (Public Domain).
Page 52, Photo: **MIN**.
Page 55, Photo: Scott Loarie, https://www.inaturalist.org/observations/33896 (CC0-1.0).
Page 56, Photo: (flower) **APP**.
Page 57, Photo: B. Semmling, https://www.inaturalist.org/observations/57477122 (CC-BY-4.0).
Page 59, Drawing: **UFL**; Photo: Darkmax, https://commons.wikimedia.org/wiki/File:Hydrilla_Verticillata_2.jpg (CC-BY-SA-3.0).
Page 60, Photo: (floating leaves) averagewalrus, https://www.inaturalist.org/observations/37138958 (CC-BY-4.0).
Page 61, Photos: bobkennedy, https://www.inaturalist.org/observations/7688231 (CC-BY-SA-4.0); A. Penney, https://www.inaturalist.org/observations/1766660 (CC-BY-4.0).
Page 62, Drawing: C.A.M. Lindman, *Bilder ur Nordens Flora* (Public Domian); Photo: W.C. Taylor @USDA-NRCS PLANTS Database/USDA NRCS. 1992. Western wetland flora: Field office guide to plant species. West Region, Sacramento (Public Domain).
Page 63, Photos: J. Lee, https://upload.wikimedia.org/wikipedia/commons/b/b4/Isoetes_echinospora_%2820262801982%29.jpg (CC-BY-SA 2.0); L. Gibson, https://www.inaturalist.org/observations/7595846 (CC-BY-4.0).
Page 64, Photo: D. Inozemtseva, https://upload.wikimedia.org/wikipedia/commons/a/a5/Isoetes_lacustris_rivnenskyi.jpg (CC-BY-SA-4.0).
Page 66, Photos: B. Armstrong, https://www.inaturalist.org/observations/15396926 (CC-BY-4.0); Q. Wiegersma, https://www.inaturalist.org/observations/59942968 (CC-BY-4.0).
Page 67, Photo: (hand) Юлия Медведько, https://www.inaturalist.org/observations/30677513 (CC-BY-4.0).
Page 68, Drawing: **WF**; Photos: Tomas P., https://www.inaturalist.org/observations/23326048 (CC0-1.0); J. Brew, https://www.inaturalist.org/observations/59348611 (CC-BY-4.0).
Page 69, Drawing: **WF**; Photo: C. Fischer, https://commons.wikimedia.org/wiki/Lemna_trisulca#/media/File:LemnaTrisulca.jpg (CC-BY-SA-3.0).
Page 70, Drawing: **WF**; Photo**:** Y. Basov, https://www.inaturalist.org/observations/50650681 (CC-BY-4.0).
Page 72, Drawing: **UFL**; Photo: K. Ozment, https://www.inaturalist.org/observations/68068875 (CC0-1.0).
Page 73, Photo: M. Bedingfield, https://canberra.naturemapr.org/Community/Sightings/Details/3390212 (CC-BY-3.0-AU).
Page 75, Photo: A. Kwiecień, Nova, https://upload.wikimedia.org/wikipedia/commons/2/2b/Ludwigia_palustris_Ludwigia_b%C5%82otna_2010-05-29_01.jpg (CC-BY-SA-4.0).
Page 76, Photo: K. Ziarnek, https://commons.wikimedia.org/wiki/File:Marsilea_quadrifolia_kz1.jpg (CC-BY-SA 4.0).
Page 78, Photos: A. Fox, University of Florida, Bugwood.org (CC-BY-3.0); mbelitz, https://www.inaturalist.org/observations/8534085 (CC0-1.0); Y. Danilevsky, https://www.inaturalist.org/observations/19902829 (CC-BY-4.0).
Page 79, Photo: R. Foster, https://www.inaturalist.org/observations/31649880 (CC-BY-4.0).
Page 80, Drawing: **UFL**; Photo: Nautilus 211, https://www.inaturalist.org/observations/42858341 (CC0-1.0).
Page 82, Drawing: **UFL**.
Page 85, Photo: © 2021 Hyun Jung Cho, http://jcho.masgc.org/.
Page 86, Photos: k2018lena, https://www.inaturalist.org/observations/27162625 (CC0-1.0); N. Filippova https://www.inaturalist.org/observations/19930673 (CC-BY-4.0).
Page 87, Drawing: **UFL**.
Page 88, Photos: S. Lefnaer, https://commons.wikimedia.org/wiki/File:Myriophyllum_spicatum_sl5.jpg (CC-BY-SA-4.0); https://commons.wikimedia.org/wiki/File:Myriophyllum_spicatum_sl1.jpg (CC-BY-SA-4.0); Елена Смирнова, https://www.inaturalist.org/observations/49493657 (CC0-1.0).
Page 89, Photo: (flowers) D. McCorquodale https://www.inaturalist.org/observations/32687637 (CC-BY-4.0).
Page 90, Photos: C. Hohn, https://www.inaturalist.org/observations/12964590 (CC-BY-4.0); Y. Basov, https://www.inaturalist.org/observations/56117341 (CC-BY-4.0).
Page 91, Drawings: **UFL**
Page 92, Photo: (*N. flexilis*) B. Isaac https://www.inaturalist.org/observations/56414038 (CC0-1.0).
Page 93, Photos: (hand) S. Johnson, https://www.inaturalist.org/observations/30429985 (CC-BY-4.0).
Page 94, Photo: R.H. Mohlenbrock, hosted by the USDA-NRCS PLANTS Database / USDA NRCS. 1992. *Western wetland flora: Field office guide to plant species*. West Region, Sacramento (Public Domain).
Page 95, Photo: Show_ryu, https://commons.wikimedia.org/wiki/File:Najas_gracillima.JPG (CC-BY-SA-3.0).
Page 96, Drawing: Gilg, E. and K. Schumann. 1900. Das Pflanzenreich Hausschatz des Wissens (Public Domain); Photo: L.J. Mehrhoff, https://www.invasive.org/browse/detail.cfm?imgnum=5446020 (CC-BY-3.0).
Page 97, Photos: https://www.inaturalist.org/observations/17986413 (CC0-1.0).
Page 98, Drawing: **UFL**; Photos: **APP**
Page 99, Photo: https://www.inaturalist.org/observations/73610980 (CC0-1.0).
Page 101, Photos: https://www.inaturalist.org/observations/23015238 (CC0 1.0). https://www.inaturalist.org/observations/6402033 (CC-BY-4.0).
Page 104, Drawing: **WF**; Photo: J. Hollinger, https://commons.wikimedia.org/wiki/File:Nymphaea_odorata_(2).jpg (CC-BY-2.0).

Page 106, Photo: (*N. peltata*) Gouwenaar, https://commons.wikimedia.org/wiki/File:18820180704_Dotterbloemen_Veersloot_Molensloot_langs_Pannepad.jpg (CC-BY-SA-4.0).
Page 108, Photo: **APP**
Page 110, Photo: (stipule) C. Hohn, https://www.inaturalist.org/observations/13809078 (CC-BY-4.0).
Page 111, Drawing: **UFL**; Photo: (hand) https://www.inaturalist.org/observations/33859843 (CC0-1.0).
Page 112, Photos: D. Goldman, hosted by the USDA-NRCS PLANTS Database (Public Domain); D. Suitor, https://www.inaturalist.org/observations/96405335 (CC-BY-4.0).
Page 114, Photos: (*P. natans*) C. Fischer, https://commons.wikimedia.org/wiki/File:PotamogetonNatans2.jpg (CC-BY-SA-3.0); (*P. gramineus*) J. Hollinger, https://commons.wikimedia.org/wiki/File:Potamogeton_gramineus_(3817604665).jpg (CC-BY-2.0); Plants (*P. obtusifolius*) K. Ziarnek, Kenraiz, https://commons.wikimedia.org/wiki/File:Potamogeton_obtusifolius_kz01.jpg (CC-BY-SA 4.0); (*P. zosteriformis*) A. Gunnar https://www.inaturalist.org/observations/15942868 (CC-BY-4.0).
Page 115, Drawings: (nodal gland) **CBH**; (plant) **UFL**; (fruit) **CBH**.
Page 118, Photos: (*P. perfoliatus*) C. Fischer, https://commons.wikimedia.org/wiki/Potamogeton_perfoliatus#/media/File:PotamogetonPerfoliatus3.jpg (CC-BY-SA -3.0); (*P. gemmiparus*) **CBH**.
Page 119, Photo: C. Hohn, https://www.inaturalist.org/observations/7825307 (CC0-1.0).
Page 120, Photos: (top) **MIN**; (Bottom) Q. Wiegersma, https://www.inaturalist.org/observations/34427597 (CC-BY-4.0).
Page 121, Drawing: K. van de Weyer & C. Schmidt, https://commons.wikimedia.org/wiki/File:Potamogeton_berchtoldii_illustration_(01).png (CC-BY-SA-3.0); Photo: Hugues TINGUY, https://commons.wikimedia.org/wiki/File:Potamogeton_berchtoldii_plant_(01).jpg (CC-BY-SA-2.0-FR).
Page 123, Drawing: UFL; Photo: N. Martineau, https://www.inaturalist.org/observations/15712219 (CC-BY-4.0).
Page 127, Photos: (hand) R. Foster, https://www.inaturalist.org/observations/20819808 (CC-BY-4.0); (fruits) C. Hohn, https://www.inaturalist.org/observations/7193394 (CC0-1.0).
Page 129, Photos: (hand) https://www.cherrug.se/nature/; (leaves) K. Peters, https://commons.wikimedia.org/wiki/File:Potamogeton_friesii.jpeg (CC-BY-SA 3.0).
Page 130, Photo: **CBH**.
Page 131, Drawing: Winterringer, *Aquatic plants of Illinois* (Public Domain); Photo: M. Bowser, https://www.inaturalist.org/observations/3642607 (CC-BY-4.0).
Page 132, Photos: (hand) R. Foster, https://www.inaturalist.org/observations/20170344 (CC-BY-4.0); (leaf) **MIN**.
Page 133, Photo: **CBH**.
Page 135, Drawing: **UFL**; Photo: **MIN**.
Page 136, Photos: S. Lefnaer, https://commons.wikimedia.org/wiki/File:Potamogeton_natans_sl22.jpg (CC-BY-SA-4.0); I. Bashinskiy, https://www.inaturalist.org/observations/88367574 (CC-BY-4.0).
Page 137, Photos: K. Ziarnek, Kenraiz, ;https://commons.wikimedia.org/wiki/File:Potamogeton_nodosus_kz04.jpg (CC-BY-SA-4.0); S. lefnaer, https://commons.wikimedia.org/wiki/File:Potamogeton_nodosus_sl3.jpg (CC-BY-SA-4.0).
Page 137, Photo: Specimen v0242029WIS, University of Wisconsin – Madison herbarium (CC0-1.0).
Page 139, Photos: A. Hosper NL https://www.inaturalist.org/observations/15521384 (CC-BY-4.0).
Page 140, Photos: K. Ziarnek, Kenraiz, https://commons.wikimedia.org/wiki/File:Potamogeton_perfoliatus_kz02.jpg (CC-BY-SA 4.0); S. Lefnaer, https://commons.wikimedia.org/wiki/File:Potamogeton_perfoliatus_sl9.jpg (CC-BY-SA-4.0); S. Lefnaer, https://commons.wikimedia.org/wiki/File:Potamogeton_perfoliatus_sl10.jpg (CC-BY-SA-4.0).
Page 141, Photos: https://commons.wikimedia.org/wiki/File:Potamogeton_praelongus,_Llyn_Cregennen_Isaf,_Wales.jpg (CC-BY-SA-4.0); K. Peters, https://commons.wikimedia.org/wiki/File:Potamogeton_praelongus.jpeg (CC-BY-SA-3.0-migrated); K. Peter, https://commons.wikimedia.org/wiki/File:Potamogeton_praelongus_bluete.jpeg (CC-BY-SA-3.0-migrated).
Page 143, Photo: S. Lefnaer, https://upload.wikimedia.org/wikipedia/commons/d/da/Potamogeton_pusillus_s._str._sl9.jpg (CC-BY-SA 4.0).
Page 144, Photos: Q. Wiegersma, https://www.inaturalist.org/observations/33154441 (CC-BY-4.0); R. Foster, https://www.inaturalist.org/observations/32722418 (CC-BY-4.0).
Page 145, Photo: M. Bowser, https://www.inaturalist.org/observations/33453024 (CC-BY-4.0).
Page 148, Photo: University of Wisconsin-Madison Herbarium, Catalog v0244303WIS (CC0-1.0).
Page 150, Photos: M. Bowser, https://www.inaturalist.org/observations/16033431 (CC-BY-4.0).
Page 151, Drawings: UFL.
Page 153, Photo: R. Watson, https://www.inaturalist.org/observations/16804436 (CC-BY-4.0).
Page 154, Photos: (*R. aquatilis*) tomopteris, https://www.inaturalist.org/observations/10004273 (CC-BY-4.0); (*R. flabellaris*) R. Martin, https://www.inaturalist.org/observations/46597959 (CC0-1.0).
Page 155, Photos: (flower) Rasbak, https://commons.wikimedia.org/wiki/File:Fijne_waterranonkel_Ranunculus_aquatilis_var._diffusus.jpg (CC-BY-SA-3.0); (leaf) garmonb0zia, https://www.inaturalist.org/observations/45565093 (CC0-1.0); (stem) APP.
Page 156, Photos: J. Hollinger, https://commons.wikimedia.org/wiki/File:Ranunculus_flabellaris_1.jpg (CC-BY-2.0); N. Martineau, https://www.inaturalist.org/observations/24767041 (CC-BY-4.0).
Page 157, Drawing: M. Cilenšek. 1892. *Naše škodljive rastline* (Public Domain); Photos: H. Rose, https://commons.wikimedia.org/wiki/File:Ranunculus_sceleratus_leaf1_(14758147920).jpg (CC-BY-2.0); bazwal, https://www.inaturalist.org/observations/18211442 (CC-BY-4.0).

Page 158, Photos:
https://commons.wikimedia.org/wiki/File:Ruppia_maritima_South_Chungcheong,_South_Korea_27_Jun_2006.jpg (CC-BY-SA-3.0); J. Horn, https://www.inaturalist.org/observations/98553556 (CC-BY-4.0).
Page 160, Photos: (*S. graminea*) https://commons.wikimedia.org/wiki/File:Grassy_Arrowhead_(Sagittaria_graminea)_(38859446382).jpg (CC-BY-2.0); (*S. cuneata*) R. Foster, https://www.inaturalist.org/observations/31359039 (CC-BY-4.0).
Page 161, Photo: **MIN**.
Page 162, Photo: Specimen 57361, University of South Florida Herbarium (CC-BY-3.0).
Page 164, Photo: (habit) I. Louque, https://www.inaturalist.org/observations/5129954 (CC-BY-4.0).
Page 166, Photos: (fruits) **MIN**; (plant) Étienne Lacroix-Carignan, https://www.inaturalist.org/observations/95723571 (CC0-1.0).
Page 167, Photo: K. Ziarnek, https://commons.wikimedia.org/wiki/File:Sagittaria_subulata_kz01.jpg (CC0-1.0).
Page 168, Photo: Specimen UTC00003032, Intermountain Herbarium, Utah State University (CC0-1.0).
Page 169, Photo: **MIN**.
Page 170, Photo: H. Hillewaert, https://commons.wikimedia.org/wiki/File:Sparganium_angustifolium.jpg (CC-BY-SA-4.0).
Page 171, Photos: M. Vainu, https://commons.wikimedia.org/wiki/Category:Sparganium_angustifolium#/media/File:Sparganium_angustifolium_in_Estonia.jpg (CC-BY-SA 4.0); https://upload.wikimedia.org/wikipedia/commons/0/05/Sparganium_angustifolium.JPG (CC-BY-SA-3.0).
Page 172, Photo: R. Routledge, https://www.invasive.org/browse/detail.cfm?imgnum=5538795& (CC-BY-3.0).
Page 173, Photos: J. Hollinger, https://commons.wikimedia.org/wiki/File:Small_Bur-Reed_(3818415894).jpg (CC-BY-2.0); A. Zharkikh, https://commons.wikimedia.org/wiki/Category:Sparganium_natans#/media/File:2016.08.27_13.16.58_IMG_7891_-_Flickr_-_andrey_zharkikh.jpg (CC-BY-2.0).
Page 174, Drawing: **UFL**; Photo: **APP**.
Page 175, Photos: C. Fischer, https://commons.wikimedia.org/wiki/File:PotamogetonPectinatus2.jpg (CC-BY-SA-3.0); S. Lefnaer, https://commons.wikimedia.org/wiki/File:Potamogeton_pectinatus_subsp._pectinatus_sl31.jpg (CC-BY-SA-4.0).
Page 177, Drawings: (*U. vulgaris*) https://commons.wikimedia.org/wiki/File:Utricularia_macrorhiza.jpg (Public Domain).
Page 178, Photos: (*U. geminiscapa*) C. Light, https://commons.wikimedia.org/wiki/File:Utricularia_geminiscapa_PinhookBog.jpg (CC-BY-SA-3.0); (*U. vulgaris*) Q. Wiegersma, https://www.inaturalist.org/observations/14405102 (CC-BY-4.0); (*U. gibba*) cwarneke https://www.inaturalist.org/observations/6744465 (CC0-1.0).
Page 179, Photos: (flowers) M. Graziano, with permission; (plant) S. Johnson, https://www.inaturalist.org/observations/30102530 (CC-BY-4.0).
Page 180, Photos: (flower) M. Graziano, with permission; (plant) R. Martin, https://www.inaturalist.org/observations/7969383 (CC0-1.0).
Page 181, Photos: (flowers) T. Eiben https://www.inaturalist.org/observations/29755043 (CC0-1.0).
Page 182, Photos: (flower) M. Graziano, with permission; (floats) B. O'Meara, https://www.inaturalist.org/observations/6850277 (CC0-1.0).
Page 183, Photo: (bladders) P. Skawinski, https://commons.wikimedia.org/wiki/File:Utricularia_intermedia_plant_(03).jpg (CC-BY-SA-3.0).
Page 184, Photo: (flower) M. Graziano, with permission.
Page 185, Photos: H. Bellat, https://commons.wikimedia.org/wiki/File:Utricularia_minor_flower_(01).JPG (CC-BY-SA-2.0-FR); Ken-ichi Ueda, https://www.inaturalist.org/observations/14710263 (CC0-1.0).
Page 186, Drawing: **UFL**; Photos: (flowers) N. Martineau, https://www.inaturalist.org/observations/15158860 (CC-BY-4.0); (plant) **MIN**.
Page 187, Drawing: **UFL**.
Page 188, Photos: (plant) **MIN**; (flower) N. Martineau, https://www.inaturalist.org/observations/15712241 (CC-BY-4.0).
Page 189, Photos: (plant) W. McFarland https://www.inaturalist.org/observations/16773557 (CC0-1.0); (flower) S. Keller https://www.inaturalist.org/observations/14302309 (CC-BY-4.0).
Page 190, Photos: (plant) N. Martineau, https://www.inaturalist.org/observations/19260491 (CC-BY-4.0); (flowers) D. Loarie, https://www.inaturalist.org/observations/6827451 (CC-BY-4.0).
Page 191, Photos: (flowers) N. Takebayashi, https://commons.wikimedia.org/wiki/File:Utricularia_macrorhiza_(29309016568).jpg (CC-BY-SA-2.0); (plant) https://www.inaturalist.org/observations/29494616 (CC0-1.0).
Page 193, Photo: J. Fortnash https://www.inaturalist.org/observations/63269150 (CC0-1.0).
Page 194, Photo: C. Fischer, https://commons.wikimedia.org/wiki/File:WolffiaArrhiza2.jpg?fastcci_from=6316484&c1=6316484&d1=15&s=200&a=fqv (CC-BY-SA-3.0).
Page 195, Photo: **APP**.
Page 196, Drawing: **UFL**; Photo: S. lefnaer, https://commons.wikimedia.org/wiki/File:Wolffia_columbiana_sl14.jpg (CC-BY-SA-4.0).
Page 197, Photos: C. Fischer, https://commons.wikimedia.org/w/index.php?curid=21435105 (CC-BY-SA-3.0).
Page 198, Photo: C. Lansdale, https://www.inaturalist.org/observations/14362436 (CC0-1.0).

GLOSSARY

Achene: A small, dry, non-opening, one-seeded fruit
Alternate: Applied to arrangement of leaves, whereby they arise singly at each node
Annual: Plants completing their life cycle in one year or growing season
Anther: The terminal portion of a stamen; bears pollen
Axil: As in leaf axil, the angle between a stem and a leaf
Basal: Applied to arrangement of leaves, whereby they are confined to the bottom or base of the plant
Bladders: small, sac-like structures specialized to capture prey
Blade: The expanded flat portion of a leaf
Bract: A much reduced leaf at the base of a flower stalk or flower cluster
Capsule: A dry fruit comprised of multiple divisions, normally opening to release seeds
Dimorphic: Having two forms, e.g., divided submersed leaves and undivided emergent leaves
Drupe: A fleshy fruit containing a seed(s) enclosed in a hard pit
Emergent: Existing above the surface of water, surrounded by air
Emersed: Projected above water surface
Flowering stem: The stalk of a flower cluster (i.e., a peduncle)
Follicle: A dry fruit which splits only along one side at maturity
Fragmentation: Process by which a plant body separates into multiple pieces, often with each detached section capable of rooting
Globose: Spherical in shape
Hydrophobic: Water-repellant
Leaf stalk: The stalk by which a leaf is attached to stem (i.e., a petiole)
Linear: Line-shaped; very long and narrow, with essentially parallel sides
Midrib: The central, linear, vein that runs along the length of a leaf blade from base to apex
Midvein: The thick central vein that runs along the length of a leaf blade
Node: A place on a stem where a leaf or leaf scar is attached
Nutlet: A diminutive nut; dry, non-opening, usually one-seeded, fruit with a hard shell
Obovate: Shaped like a section through a hen's egg but with the larger end toward the top
Opposite: Applied to arrangement of leaves, buds, or scars whereby they arise in pairs positioned across from each other, one pair at each node
Ovary: The usually swollen base of the female organ which contains the eventual seeds
Palmately divided: Arrangement of leaf segments or lobes, or veins, with these diverging from a common origin at the top of a leaf stalk
Perennial: A plant that lasts for more than two years
Petal: One of the inner floral leaves in a flower, usually brightly colored, which are whorled
Phyllodial: The condition of a leaf whereby the leaf stalk is expanded and functions as a leaf, i.e., a bladeless leaf
Pinnately divided: Arrangement of leaf segments or lobes, or veins, with these displayed on either side of a central stalk or midrib, respectively
Receptacle: The expanded base of a flower
Recurved: Curved or bent downward
Rhizome: A horizontal underground stem with roots growing from it
Rosette: A circular cluster of tightly packed leaves, usually at ground level
Sepal: One of the outer floral leaves in a flower, usually greenish, which are whorled
Septate: With internal transverse partitions or walls called septa
Sinus: The space between two lobes of a leaf
Spike: An elongated, unbranched stem bearing flowers along it that

lack stalks; youngest flowers at top

Spikelet: A very small spike consisting of several reduced flowers concealed behind scales; mostly in grasses and sedges

Stamen: The male, pollen-producing, organ of a flower comprising a stalk and the anther

Stipule: A leafy or linear structure at or near the base of the leaf stalk

Stolon: A horizontal aboveground stem with roots growing from it

Stomate: A leaf pore used for gas exchange

Submersed: Existing under the surface of water

Two-ranked: Where the leaves on a stem are arranged in two vertical columns on opposite sides of the stem, appearing as two rows of leaves

Tuber: A thickened portion of an underground stem serving as a storage organ, e.g., a potato

Tubercle: A low, round bump, e.g., on top of some *Eleocharis* fruits

Turion: A highly condensed vegetative organ specialized for overwintering, i.e., winterbud

Whorled: Applied to arrangement of leaves or flowers whereby they arise 3 or more at each node

Winterbud: See *turion*

INDEX

ABOUT THE AUTHOR

Donald J. Padgett, Professor of Biological Sciences, holds a B.S. degree in biology from Susquehanna University, Selinsgrove, Pennsylvania, and an M.S. and Ph.D. in plant biology from the University of New Hampshire, Durham. He teaches wetland ecology, invasion ecology, and biology at Bridgewater State University, Bridgewater, Massachusetts. His research focuses on the systematics of aquatic and wetland plants. He is also the author of *Aquatic Plants of Massachusetts* and *Wetland Plants of New England: A Guide to Trees, Shrubs, and Lianas*.

Made in the USA
Las Vegas, NV
03 November 2026

57423533R00133